Introduction to

MAGNETOCHEMISTRY

Introduction to

MAGNETOCHEMISTRY

ALAN EARNSHAW

*Department of Inorganic and Structural Chemistry
University of Leeds, England*

1968

ACADEMIC PRESS

LONDON and NEW YORK

ACADEMIC PRESS INC. (LONDON) LTD.
Berkeley Square House
Berkeley Square
London, W.1

U.S. Edition published by
ACADEMIC PRESS INC.
111 Fifth Avenue
New York, New York 10003

Library of Congress Catalog Card Number: 67-30769

Printed in Great Britain by
Alden & Mowbray Ltd
at the Alden Press, Oxford

PREFACE

Of the many techniques available to co-ordination chemists, the measurement of magnetic moment has been one of the most consistently useful. For teaching purposes it provides a simple method of illustrating the ideas of electronic structure, and in research it can provide fundamental information about the bonding and stereochemistry of complexes.

A number of reviews and advanced treatises have recently been published but, since the excellent book by Selwood (see group 13, p. 112), little has appeared that would give the beginner in the subject a general outline of the experimental techniques along with an account of the sort of chemical information which they yield. The object of this book is to provide an introduction to the more important aspects of magnetochemistry. The treatment is certainly not intended to be exhaustive and detailed references have been omitted, although the way into the original literature can be found through the more comprehensive reviews which are cited.

Of the various approaches which have been used to explain the magnetic properties of co-ordination compounds, the most fruitful is that based on a consideration of the effect of ligands on the spectroscopic terms of metal ions. Most students of chemistry start with a background understanding of their subject based on the concept of orbitals of definite shapes and energies, and it is the electron occupancy of these orbitals which is the basis of Valence Bond and simple Crystal Field theories. Unfortunately the transition from these to the more abstract consideration of spectral terms is often the cause of appreciable difficulty to the more practically minded student. This is primarily a conceptional difficulty and seems to be most easily overcome by the use of "pictorial" explanations. These are therefore used freely even, on occasions, at the expense of strict accuracy. There are obvious risks in this approach but I believe they are justified providing the limitations are made plain.

It is a pleasure to extend my thanks to the many friends who have assisted in the preparation of this book: to Professor J. Lewis for reading the manuscript and making a number of helpful comments; to the students and lecturers, particularly Dr L. F. Larkworthy, of the University of Surrey where most of the writing was done, for many helpful discussions, and to Mrs Pam Spicer who typed the manuscript.

January 1968

ALAN EARNSHAW

CONTENTS

DEFINITIONS OF SYMBOLS

Symbol	Definition	Page Introduced
w	change in weight	85
W	weight	85
α	a constant	58
	used also in $N\alpha$	27
β	Bohr magneton	4
γ	type of d orbital	38
Δ	crystal field splitting	38
ε	type of d orbital	38
ζ	spin-orbit coupling constant for an electron	59
θ	Curie-Weiss (or Weiss) constant	3
κ	volume magnetic susceptibility	1
λ	constitutive correction	6
	spin-orbit coupling constant for a term	58
μ	magnetic moment	4
ρ	density	12
χ	magnetic susceptibility	1
ω	angular velocity	9

ABBREVIATIONS

acac	acetylacetonate
bipy	bipyridyl
DMGH	dimethylglyoxime
diarsine	o-phenylenebisdimethylarsine
en	ethylenediamine
phen	phenanthroline
py	pyridine
salen	salicylaldehyde-ethylenediimine

I. INTRODUCTION

Since the time of Faraday it has been realized that all substances possess magnetic properties. That is to say, all substances are affected in some way by the application of a magnetic field. The general expression of this fact is as follows.

If a substance is placed in a field of H oersteds then B, the magnetic induction, or the density of lines of force within the substance, is given by H plus a contribution, $4\pi I$, due to the substance itself.

$$B = H + 4\pi I \tag{1}$$

where I is the intensity of magnetization, or magnetic moment per unit volume. Dividing equation (1) by H:

$$P = 1 + 4\pi\kappa \tag{2}$$

where P and κ are the permeability and susceptibility per unit volume respectively and which may be considered dimensionless.

In practice, susceptibility is usually more conveniently expressed per unit mass (gram susceptibility) than per unit volume.

$$\chi = \frac{\kappa}{\text{density}}$$

The molar susceptibility, designated χ_M, is then $\chi \times$ mol. wt.

Equation (2) leads to the most fundamental magnetic classification of substances

1. $P < 1$ i.e. I, κ and χ negative

In this case the substance is said to be *diamagnetic* and causes a reduction in the density of lines of force. Since this is equivalent to the substance producing a flux opposed to the field causing it, it follows that, if the field is non-homogeneous, the substance will tend to move to regions of lowest field strength. Experimental values of χ are negative and are found to be very small

($\simeq -1 \times 10^{-6}$) and generally independent both of field strength and temperature.

2. $P > 1$ i.e. I, κ and χ positive

In this case the substance is said to be *paramagnetic* and causes an increase in the density of lines of force. This is equivalent to the substance producing a flux in the same direction as the field. Thus, in a non-homogeneous field, the substance will tend to move to regions of highest field strength. Experimental

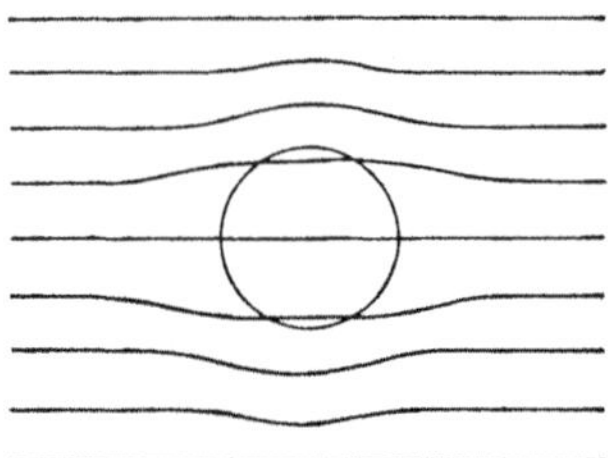

Fig. 1. Diamagnetic body in a magnetic field.

values of χ are positive and are found to be rather larger than in the diamagnetic case (1 to 100×10^{-6}). Though independent of field strength, χ is markedly dependent on temperature.

Ferromagnetic and *antiferromagnetic* materials are subdivisions of this class. In the former case $P \gg 1$ and χ can be as high as 10^4, being both field and temperature dependent. Remanence and hysteresis are characteristic properties of ferromagnetic materials. In the latter case, χ is usually somewhat

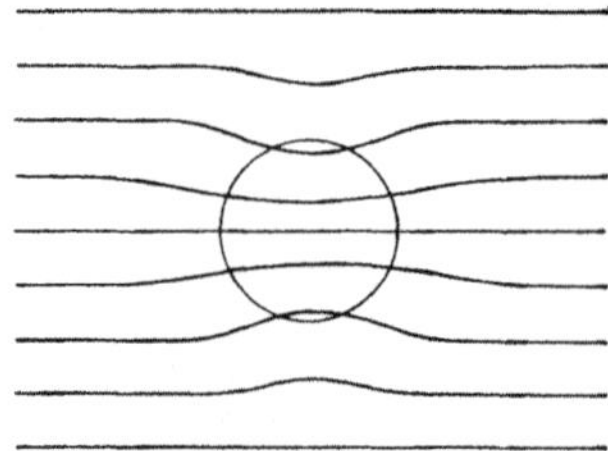

Fig. 2. Paramagnetic body in a magnetic field.

less than for normal paramagnetic materials and is temperature and, sometimes also, field dependent. Ferro- and antiferromagnetism are known as "co-operative phenomena". That is to say they arise when the paramagnetic centres within a sample of the substance interact magnetically with each other. They are thus properties of materials in bulk rather than of the individual atoms or ions, and are more likely to occur when there is a high concentration of paramagnetic centres in the substance. Arising from this is the rather arbitrary, but none the less useful, distinction between magnetically concentrated and magnetically dilute materials. The former are exemplified by pure metals and alloys and also by compounds such as oxides and fluorides in which the metal ions are combined with small anions. Examples of the latter are solutions of paramagnetic compounds and solids in which a metal ion is co-ordinated to large ligands or is heavily hydrated. The situation is complicated by the fact that, even in compounds which are apparently magnetically dilute, interactions between the paramagnetic centres may occur if the molecular or crystal structure is such as to bring them sufficiently close or if they are joined by suitable bridging groups. This happens in many binuclear and polynuclear complexes which will be dealt with later.

At the end of the nineteenth century Curie, while investigating the effect of temperature on magnetic properties, found that, for a number of paramagnetic substances, χ and T are inversely proportional

$$\chi = \frac{C}{T} \tag{3}$$

This is the well-known Curie law and C is the Curie constant for the substance in question.

In general, paramagnetic substances will be composed of paramagnetic centres and diamagnetic groups for which a correction must be applied (indeed, as will be seen later, even a monatomic paramagnetic ion has an underlying diamagnetism). Thus the molar susceptibility of a substance is the algebraic sum of the susceptibilities of the component atoms, ions or molecules. The susceptibility, per gram atom, χ_A of, say, a paramagnetic metal ion in a particular compound can therefore be obtained by measuring the molar susceptibility of the compound and subtracting from this the diamagnetism of the other ions or molecules in the compound.

Although Curie originally formulated his law in terms of χ, it is nowadays usually expressed in terms of χ_A.

Curie also discovered that for every ferromagnetic substance there is a temperature (the Curie temperature or point) above which normal paramagnetic behaviour occurs.

Later work by Onnes and Perrier showed that, for many paramagnetic substances, a more exact relationship is

$$\chi_A = \frac{C}{T+\theta} \tag{4}$$

Weiss was able to justify this theoretically and, although his explanation is now realized to be less widely applicable than was originally thought, the equation is known as the Curie-Weiss law; θ being the Curie-Weiss constant. θ is sometimes referred to as the Curie temperature since, in the case of some ferromagnetic materials at temperatures above their Curie points, the value of

θ, as predicted by Weiss, is indeed equal to the negative of their Curie temperature. For this reason also the law is sometimes given in the form

$$\chi_A = \frac{C}{T-\theta}$$

This convention is adopted by the German and Indian schools of magnetochemistry but the British and American schools use the convention of equation (4), which will be used in the present work. On this basis, when the Curie-Weiss law applies, a plot of $1/\chi$ against T is a straight line with an intercept of $-\theta$ on the T axis.

The properties of a paramagnetic material are frequently more conveniently expressed in terms of μ_e, the "effective magnetic moment". As will be seen later, this is related to the atomic susceptibility of the paramagnetic centre by the equation

$$\chi_A = \frac{N\beta^2 \mu_e^2}{3kT} \tag{5}$$

where N is Avogadro's number $= 6 \cdot 023 \times 10^{23}$

β is the Bohr magneton $= 0 \cdot 9273 \times 10^{-20}$ erg/gauss

k is Boltzmann's constant $= 1 \cdot 381 \times 10^{-16}$ erg/deg.

T is the temperature in °K

$\therefore \quad \mu_e = 2 \cdot 828 \sqrt{\chi_A T}$ Bohr magnetons

It is common practice to indicate the temperature at which a measurement is made by the addition of a subscript. For instance, μ_{300} is the effective magnetic moment at a temperature of 300°K.

In the belief that all magnetic moments were simple multiples of a common unit, Weiss introduced the Weiss magneton which was used in much of the early literature. 1 B.M. $= 4 \cdot 97$ W.M. No theoretical significance now attaches to this.

DIAMAGNETIC CORRECTIONS

As was mentioned above, χ_A for a paramagnetic ion may be obtained by subtracting from the molar susceptibility the susceptibilities of the diamagnetic groups or ligands also present

$$\chi_A = \chi_M - \sum \chi_L$$

The diamagnetism of molecules can be measured directly and various methods have been suggested for determining the susceptibilities of diamagnetic ions. None of these is entirely satisfactory. Perhaps the simplest is to start with the assumption that the hydrogen ion, having no electrons, has zero susceptibility. Hence the susceptibility of an acid is that of its anion. The main objection to this is that the hydrogen ion, because of its small size, exerts a distorting effect on neighbouring ions which will not occur to the same extent

TABLE I. *Molar susceptibilities (χ_L) of common ligands and ions*
All values $\times 10^{-6}$/mole

Cations		Anions	
Li^+	-1.0	F^-	-9.1
Na^+	-6.8	Cl^-	-23.4
K^+	-14.9	Br^-	-34.6
Rb^+	-22.5	I^-	-50.6
Cs^+	-35.0	CN^-	-13.0
NH_4^+	-13.3	CNS^-	-31.0
Mg^{2+}	-5.0	CO_3^{2-}	-28.0
Ca^{2+}	-10.4	ClO_4^-	-32.0
Zn^{2+}	-15.0	NO_2^-	-10.0
Hg^{2+}	-40.0	NO_3^-	-18.9
		OH^-	-12.0
		O^{2-}	-7.0
		$PtCl_6^{2-}$	-148
		SO_4^{2-}	-40.1
		SO_4H^-	-35.0

Common ligands

H_2O	water	-13	$C_2O_4^{2-}$	oxalate	-25
NH_3	ammonia	-18	$C_3H_2O_4^{2-}$	malonate	-45
N_2H_4	hydrazine	-20	$C_4H_7O_2^-$	acac	-52
CO	carbonyl	-10	$C_5H_5^-$	cyclopenta-	
CHO_2^-	formate	-17		dienyl	-65
CH_4N_2O	urea	-34	C_5H_5N	py	-49
CH_4N_2S	thiourea	-42	$C_9H_6NO^-$	oxinate	-86
C_2H_4	ethylene	-15	$C_{10}H_8N_2$	bipy	-105
$C_2H_3O_2^-$	acetate	-30	$C_{12}H_8N_2$	phen	-128
$C_2H_4NO_2^-$	glycinate	-37	$C_{16}H_{14}N_2O_2^{2-}$	salen	-182
$C_2H_8N_2$	en	-46	$C_{10}H_{16}As_2$	diarsine	-194
			$C_{32}H_{16}N_4^{2-}$	phthalo-	
				cyanine	-422

in other compounds of these ions. It should also be noted that, although these diamagnetic susceptibilities are additive* for a particular compound, they will not necessarily be constant in different compounds. Factors such as the polarization of the ion, which varies from compound to compound, will affect

* We are considering here ionic or molar susceptibilities. Gram susceptibilities are also additive providing allowance is made for the weight composition of the substance. $\chi = \chi_1 p_1 + \chi_2 p_2 + \ldots \chi_n p_n$ where χ_n is the gram susceptibility and p_n the weight fraction of a particular component. This is known as Wiedemann's law.

the diamagnetism. In practice, however, it is possible to assign average values to these diamagnetic susceptibilities which, though subject to quite high percentage errors, are acceptable since the values themselves are considerably smaller than the paramagnetic susceptibilities to which they are applied as corrections.

Table I gives values of the susceptibilities of a number of the more commonly occurring ions and ligands.

TABLE II. Pascal's constants (χ_A)
All values $\times 10^{-6}$/g atom.

H		$-2\cdot93$	O_2	(carboxylate)	$-7\cdot95$
C		$-6\cdot00$	F		$-6\cdot3$
N		$-5\cdot57$	Cl		$-20\cdot1$
N	(ring)	$-4\cdot61$	Br		$-30\cdot6$
N	(monamide)	$-1\cdot54$	I		$-44\cdot6$
N	(diamide, imide)	$-2\cdot11$	S		$-15\cdot0$
O		$-4\cdot61$	P		$-26\cdot3$
O	(ketone)	$+1\cdot73$	As(III)		$-20\cdot9$

There is no doubt that wherever possible diamagnetic molar susceptibilities should be measured directly. However, it often happens that this is highly inconvenient and, in such cases, recourse may be made to Pascal's constants from which these susceptibilities may be calculated.

TABLE III. Constitutive corrections (λ)
All values $\times 10^{-6}$.

C in ring	$-0\cdot24$	C=N	$+8\cdot15$
C shared by two rings	$-3\cdot07$	C≡N	$+0\cdot8$
C=C bond	$+5\cdot5$	N=N	$+1\cdot8$
C≡C bond	$+0\cdot8$	N=O	$+1\cdot7$

As a result of measurements on a wide variety of compounds, Pascal concluded that molar susceptibility could be expressed as

$$\chi_M = \sum n_A \chi_A + \sum \lambda$$

where n_A is the number of A atoms of atomic susceptibilities χ_A in the molecule; λ is a "constitutive correction" which depends on the nature of the bonds involved, and takes account of variations in χ_A which depend on the different environments of the atom in various compounds.

The values of χ_A are thus intended to be specific to the particular atoms irrespective of their environments and were deduced empirically. Taking the atomic susceptibility of Cl as half the molar susceptibility of Cl_2, Pascal calculated the atomic susceptibility of H by replacing Cl by H in a series of monochloro hydrocarbons. Table II is a collection of atomic susceptibilities (Pascal's constants) calculated similarly, and Table III gives some of the constitutive corrections.

The actual values obtained by Pascal were based on a value for the susceptibility of his calibrant, water, which is now realized to have been in error. All the above values have been corrected for this.

The use of Pascal's constants can be illustrated by the following examples:

Pyridine, C_5H_5N

$$5 \times C = -30{\cdot}00 \times 10^{-6}$$
$$5 \times H = -14{\cdot}65 \qquad \sum \lambda = 5 \times \text{ring C} = -1{\cdot}2 \times 10^{-6}$$
$$\text{ring N} = -\ 4{\cdot}61$$
$$\therefore \quad \sum \chi_A = -49{\cdot}26 \times 10^{-6}$$

$\therefore$ total $\simeq -50{\cdot}5 \times 10^{-6}$ compared to the experimental value of -49×10^{-6}

Acetate, $C_2H_3O_2{}^-$

$$2 \times C = -12{\cdot}00 \times 10^{-6}$$
$$3 \times H = -\ 8{\cdot}79$$
$$\text{carboxylate } O_2 = -\ 7{\cdot}95 \qquad \sum \lambda = 0$$
$$\therefore \quad \sum \chi_A = -28{\cdot}74$$

$\therefore$ total $\simeq -28{\cdot}7 \times 10^{-6}$ compared to the experimental value of -30×10^{-6}

Phenanthroline $C_{12}H_8N_2$

$$12 \times C = -72{\cdot}00 \times 10^{-6} \qquad 8 \times \text{ring C} = -\ 1{\cdot}9$$
$$8 \times H = -23{\cdot}44 \qquad 4 \times \text{binuclear C} = -12{\cdot}3$$
$$2 \times \text{ring N} = -\ 9{\cdot}22 \qquad \sum \lambda = -14{\cdot}2 \times 10^{-6}$$
$$\therefore \quad \sum \chi_A = -104{\cdot}66 \times 10^{-6}$$

$\therefore$ total $\simeq -118{\cdot}9 \times 10^{-6}$ compared to the experimental value of -128×10^{-6}

These examples demonstrate the sort of accuracy which can be expected. The method is satisfactory for simple molecules but the errors are greater for more complicated systems, particularly if they possess appreciable aromatic character. Far greater precision can be obtained if measurements are made on a series of similar compounds so that better values of λ can be derived. In this way it has been possible to distinguish between isomers (e.g. keto-enol isomers) by comparing experimentally determined susceptibilities with values calculated for the different isomers. However, the number of measurements needed to obtain sufficiently accurate λ values can be extremely large and while, in organic systems, this is very irksome, in inorganic systems it is rarely possible even to obtain the requisite series of similar compounds. Various attempts have been made to improve and simplify this method. Since the bulk of the diamagnetism of a molecule originates from the bonding rather than the non-bonding electrons, it has been suggested that a system of bond increments would be preferable to atomic increments. Quantum-mechanical methods have been used to estimate the diamagnetism of bonding electrons and also non-bonding inner electrons, and precise measurements of diamagnetic susceptibilities are extremely useful in testing the approximations used. However, the inorganic chemist is primarily concerned with diamagnetic susceptibilities only as corrections to be applied to paramagnetic compounds and for this purpose the original system of Pascal is still most widely employed. The errors involved may be high compared to the actual correction, but are usually low compared to the paramagnetic susceptibilities which are being measured. None the less it is necessary to be aware of this possible error and, as mentioned previously, direct measurements on more complicated ligands are to be preferred.*

* For references for this chapter, see groups 8 and 13 on p. 112.

II. FREE ATOMS AND IONS

Many electronic theories of magnetism have been developed since Weber in the mid-nineteenth century showed, on the basis of Amperian currents circulating within molecules, that diamagnetism should be a property common to all matter. Langevin in 1905 produced the classical theories of diamagnetism and paramagnetism which underwent modification after the application, by Bohr, of the concept of quantization to the theory of the atom. More recent quantum-mechanical treatments by Van Vleck and others have provided a more comprehensive theory of the subject.

While it is outside the scope of this book to attempt a rigorous mathematical treatment, many results can be satisfactorily derived by the use of the classical model introducing, where necessary, the results of later theories. A convenient starting point is a consideration of a single electron system from which the basic connexion between angular momentum, whether orbital or spin, and magnetic moment can be derived. From this basis it is then possible to extend the treatment to more practical, multi-electron systems.

SINGLE ELECTRON SYSTEM

For simplicity the atom or ion may be represented by the solar model in which the electron is considered to rotate about a positively charged nucleus.

THE ORBITAL MOMENT

If a given electron rotates with an average angular velocity ω, it is equivalent to a current of $e\omega/2\pi c$ e.m.u. Such a current will, of course, produce a magnetic field perpendicular to the plane of the rotation, the magnetic moment of which will be given by the product of the current and the area of the orbit

$$\text{moment} = \frac{e\omega}{2\pi c} \cdot \pi \bar{r^2}$$

i.e.
$$\mu_l = \frac{e\omega \overline{r^2}}{2c} \tag{1}$$

where $\overline{r^2}$ is the mean square radius of the orbit. Now the angular momentum of the electron is $m\omega r^2$ and, as shown by Bohr, it is quantized in units of $h/2\pi$. The wave mechanical treatment of the atom shows that its actual magnitude in terms of the orbital quantum number, l, is $\sqrt{l(l+1)}h/2\pi$. (It is in fact a general result of wave mechanics that angular momentum is always related to the quantum number in this way.)

$$\therefore \quad m\omega r^2 = \sqrt{l(l+1)}\,\frac{h}{2\pi}$$

$$\therefore \quad \omega r^2 = \sqrt{l(l+1)}\,\frac{h}{2\pi m} \tag{2}$$

Combining equations (1) and (2)

$$\mu_l = \frac{eh}{4\pi mc}\sqrt{l(l+1)}$$

The factor $eh/4\pi mc$ is given the symbol β and is termed the "Bohr magneton", the unit in which magnetic moments are conventionally given

$$\mu_l = \beta\sqrt{l(l+1)} \text{ e.m.u.}$$

Substituting appropriate values gives $\beta = 0 \cdot 9273 \times 10^{-20}$ ergs gauss^{-1}.

The Spin Moment

In order to explain certain spectral phenomena Uhlenbeck and Goudsmit in 1926 postulated that an electron possesses angular momentum in addition to the orbital angular momentum just described. This they ascribed to the electron spinning about its own axis as well as rotating about the nucleus.

This concept of electron spin should not be used too literally. That the electron possesses quantized angular momentum additional to the orbital angular momentum is not in doubt, but the physical picture of a spinning sphere is to be regarded purely as a conceptual aid. The further suggestion was also made that, for spin, the ratio of magnetic moment to angular momentum is twice that for orbital angular momentum. This assumption is essential in explaining a host of experimental data and must be regarded as correct. A rigorous treatment is provided by the relativistic wave mechanics of Dirac from which the spin quantum number and the magnetic anomaly of spin appear as necessary results.

Remembering again the wave mechanical result that spin angular momentum is not $s\,h/2\pi$ but $\sqrt{s(s+1)}\,h/2\pi$, it follows that the spin moment is

$$\mu_s = g\beta\sqrt{s(s+1)} \text{ e.m.u.} \tag{3}$$

where g is the "splitting factor", often called simply, the "g factor". It is the ratio of the magnetic moment to angular momentum where both quantities are expressed in their respective quantum units. Its exact value in this case is $2 \cdot 002320 \pm 0 \cdot 000004$. It follows that, to a very good approximation, the orbital g value is 1 and the spin g value is 2.

We have yet to discuss the way in which the spin and orbital moments of the individual electrons within a poly-electronic atom will interact with each other. However, it is evident that, because they are permanent moments, they will interact with an externally applied field to produce a paramagnetic effect. Like small bar magnets, they will tend to align themselves, or their resultant will tend to align itself, with the applied field, thus reinforcing it. Before proceeding with this, however, we must first consider the possible induced effects which an applied field may have on our atomic model.

DIAMAGNETISM

Using again the solar model and taking the general case in which the plane of the electron's orbit is inclined at some angle to the direction of the applied field, it will be seen that the field exerts a force on the electron. The force is such as to produce a torque on the orbit which causes it to precess about the field direction.

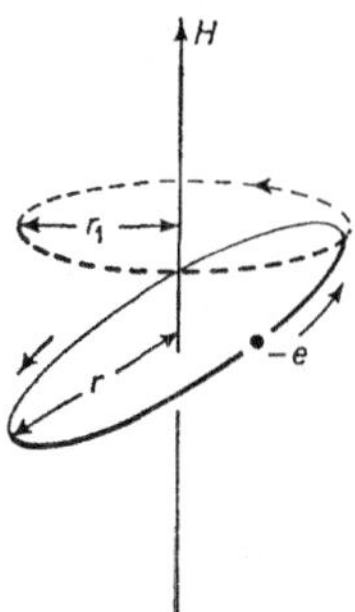

FIG. 3. Lamor precession of an electron orbit about the direction of H.

Lamor showed that this precession has an angular velocity of $eH/2mc$. It is equivalent to an extra rotation of charge about the nucleus, superimposed on the orbital motion already discussed. As such it will generate a magnetic field and the direction of this is found to be opposite to the direction of H. This conforms to Lenz's law that the direction of induced currents is such as to oppose the field causing them. It is thus a diamagnetic effect.

By analogy with the straightforward electron rotation equation (1), the moment $\Delta\mu$ produced by the Lamor precession is

$$\Delta\mu = -\frac{e^2 H}{4mc^2} \overline{r_1^2} \tag{4}$$

where $\overline{r_1^2}$ is now the mean square radius of the projection of the orbit perpendicular to the direction of H, and the negative sign is indicative of the direction of $\Delta\mu$.

Since paramagnetism rather than diamagnetism is of chief interest to chemists, it may perhaps be as well to complete the theoretical discussion of diamagnetism at this point by considering the extension to a multi-electron atom and the derivation of the atomic susceptibility.

Let the co-ordinates of the electron at a given instant be x, y, z with the nucleus as origin and the Z axis coincident with the direction of H. Then for a time average:

$$\overline{r^2} = \overline{x^2+y^2+z^2}$$

Taking the projection of the orbit $\perp$ to H (i.e. in the XY plane) and assuming all possible orientations,

$$\overline{r_1^2} = \overline{x^2+y^2} = \tfrac{2}{3}\overline{r^2}$$

$$\therefore \quad \Delta\mu = -\frac{e^2H}{6\pi mc^2}\,\overline{r^2}$$

For a multi-electron atom

$$\Delta\mu = -\frac{e^2H}{6\pi mc^2}\sum \overline{r^2}$$

and for a gram atom

$$\Delta\mu_A = -N\frac{e^2H}{6\pi mc^2}\sum \overline{r^2}$$

It will be recalled from Chapter I that I is the magnetic moment per unit volume, i.e.

$$I = \frac{\Delta\mu_A}{A}\rho$$

where ρ is the density and A the atomic weight, also the susceptibility per gram atom is

$$\chi_A = \frac{\kappa}{\rho}\,A = \frac{I}{H\rho}\,A$$

$$\therefore \quad \chi_A = \frac{\Delta\mu_A}{H}$$

$\therefore$ from equation (4)
$$\chi_A = -N\frac{e^2}{6\pi mc^2}\sum \overline{r^2} \tag{5}$$

It is comforting to note that the quantum mechanical approach of Van Vleck yields precisely the same result.

Equation (5) should only apply strictly to monatomic gases, additional terms being needed for polyatomic gases and for solids. Nevertheless some interesting points emerge from this expression. In the first place the susceptibility, being proportional only to $\sum \overline{r^2}$, should be independent of H and T. This is usually found experimentally to be a surprisingly accurate assessment of the situation even for solids. Small variations are, of course, common but can easily be accounted for on the basis of slight changes in the size of the electronic orbits. Large variations are much rarer and generally occur, as in the case of bismuth, at particular temperatures.

Measurement of χ_A and insertion in equation (5) of numerical values for the constants N, e, m and c lead to values of r of the order of 10^{-8} cm. This can only be a rough average for all the electrons in a given atom, but the fact that it is of the correct order of magnitude is indicative of the inherent applicability of equation (5).

Finally it may be emphasized that whereas paramagnetic effects are caused in the main by permanent magnetic dipoles within the atom, diamagnetism is a purely induced effect. Thus, while situations are easily visualized in which the permanent moments within a multi-electron atom or molecule cancel each other, the diamagnetic effect will be a universal feature of all matter even if masked by overlying paramagnetism.

MULTI-ELECTRON SYSTEMS

The fundamental connexion between angular momentum and paramagnetism has already been noted. In order to determine the paramagnetic effect in systems possessing more than one electron, it is necessary to deduce the resultant angular momentum for all the electrons. This is a problem with which atomic spectroscopists have long been familiar and which is conveniently approached by considering the various types of interaction, or coupling, which are possible between the electrons of an atom or ion.

Before proceeding with this problem the relationship between quantum numbers and angular momenta must be explained.

The principal quantum number, n, defines approximately the extent of the region in which the electron is most likely to be situated. It can take any integral value $(0, 1, 2 \ldots \infty)$, each value corresponding to a widely different energy.

l is the orbital quantum number and can take any integral value from 0 to $(n-1)$. It describes the shape of an orbital and also partly determines its energy, though it is not in general so important as n in this respect. A particular value of l is associated with an *orbital* angular momentum vector, l, where $l = \sqrt{l(l+1)}$ in units of $h/2\pi$. This vector may take up $2l$ different orientations in space. If a particular direction in space is defined, for instance by the application of an external magnetic field, the projections of l in this direction range from $+l$ to $-l$ and are known as the magnetic quantum numbers m_l (Fig. 4). It is a consequence of Heisenberg's uncertainty principle that the projection of l in the direction of the field cannot equal $\sqrt{l(l+1)}$.

In a similar manner the spin quantum number, s, is associated with a spin angular momentum vector, s, where $s = \sqrt{s(s+1)}$ in units of $h/2\pi$. (In the ensuing discussion angular momentum vectors will all be written in bold type, s, l, etc.) The value of s is always $\frac{1}{2}$ and s can only be parallel or anti-parallel to a magnetic field (such a field is not necessarily externally applied, it may be the field associated with l). The projection of s in the direction of this field is therefore $+s$ or $-s$, i.e. $m_s = \pm\frac{1}{2}$. As with l, the projection can never equal $\sqrt{s(s+1)}$.

The spin and orbital angular momentum vectors of electrons are able to couple together in a variety of ways which correspond to the different possible interactions. These interactions may be exchange, electrostatic, or magnetic in origin and the effect of quantization is that in each case a number of discrete arrangements result, corresponding to different energies. For each particular energy the individual angular momentum vectors assume a particular inclination to each other and precess about the direction of their resultant

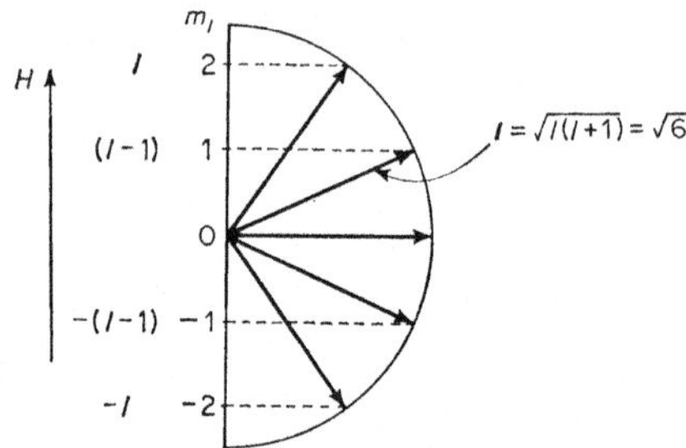

FIG. 4. The origin of magnetic quantum numbers as the projections of l in the direction of H.

as predicted by the Larmor theorem. The vectors are then said to have coupled and the angular velocity of the precession is proportional to the energy associated with the arrangement. Such coupling is usually represented by vectorial diagrams, and this practice will be followed here. However, in quantum mechanics angular momentum vectors are actually matrices, so it should be no surprise if simple geometrical diagrams prove, occasionally, to be somewhat inadequate. These diagrams are most helpful if they are understood to be merely a convenient form of representation.

The types of coupling which can occur may be represented as

$$s_i s_k, \quad l_i l_k, \quad s_i l_i \quad \text{and} \quad s_i l_k$$

where i and k signify different electrons.

$s_i s_k$ OR SPIN COUPLING

This is the coupling of the spin angular momentum vectors of different electrons. Due to the Heisenberg exchange phenomenon, different energies are associated with different orientations of these vectors. This exchange effect is actually orbital in nature but in fact the greatest exchange stabilization is possible only when the electron spins are parallel, i.e. when the resultant spin angular momentum is a maximum. This is the explanation of Hund's first rule.

$l_i \, l_k$ OR ORBITAL COUPLING

This is the coupling of the orbital angular momenta of different electrons at discrete angles to each other. The vectors (represented by l) precess about a common axis while maintaining their mutual inclination, and give rise to a resultant vector (represented by L) along this axis. Because the angular momentum vector of an electron is perpendicular to the plane of the electron's rotational motion, it follows that the planes of the orbits themselves are inclined to each other. The reason why different inclinations correspond to different energies can be seen by considering the electrostatic repulsions between electrons in these inclined orbits.

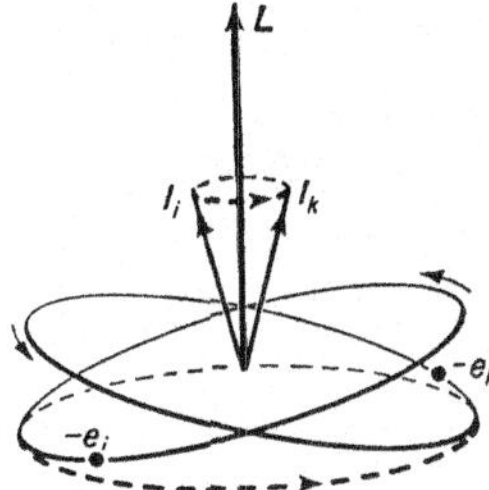

FIG. 5. Electron orbits and the associated l vectors precessing about the direction of the resultant L.

In order to keep electrostatic repulsions at a minimum, electrons will tend to occupy positions on opposite sides of the nucleus where they are as far away from each other as possible. Such relative positions can be maintained if the electrons rotate about the nucleus in the same direction. Any departure from this must increase the interelectronic repulsions and hence the energy of the system. Thus, within the restrictions of the exclusion principle, the electrons will tend to situate themselves such that the planes and directions of rotation are most nearly identical. In other words the most stable arrangement is that in which the l vectors are as nearly aligned as possible, giving rise to the maximum possible resultant. This is the explanation of Hund's second rule.

$s_i \, l_i$ OR SPIN-ORBIT COUPLING

This is the coupling of the spin and orbital angular momenta of the same electron. Again the vectors (represented by s and l) are inclined at fixed angles to each other and precess about the direction of their resultant (represented by j).

The interaction causing the vectors to couple in this way is magnetic in origin and is best visualized from the position of the individual electron. Its rotation around the nucleus then appears as a rotation of the nucleus around the electron. This produces a magnetic field, dependent on the effective charge of the nucleus, which interacts with the magnetic moment due to the spin of the electron. "Effective" must be stipulated because, of course, the nucleus will in general be screened by other electrons.

It is evident that the strength of spin-orbit coupling will tend to increase for the heavier elements because the nuclear charge is greater, and also in cases where the electron penetrates close to the nucleus, so reducing the screening effect of other electrons.

$s_i\, l_k$ Coupling

This is also a form of spin-orbit coupling but occurs between the spin and orbital angular momenta of different electrons. It is, however, too small to be of any real consequence.

In all the above cases of precessing vectors, it should be noted that, when the coupling is strong and the precession is very rapid, the individual components cease to have any significance and only the resultant vector has an exact meaning.

The resultant angular momentum for the whole atom or ion will depend on the relative magnitudes of the above interactions. If any of the interactions are of comparable magnitude the situation is very complicated, but two extreme situations may arise:

$$1. \quad s_i s_k > l_i l_k > s_i l_i$$
$$2. \quad s_i l_i > s_i s_k, \quad l_i l_k$$

Case 1 is described by the Russell-Saunders, or LS, coupling scheme which is found to apply to the lighter elements up to the end of the first transition series, and to the lower energy levels of the heavier elements. Case 2 is described by the jj coupling scheme which, although it seldom occurs pure, holds partially for heavier elements, particularly in their excited states.

Russell-Saunders, or LS, Coupling

The strongest coupling of vectors will naturally take precedence over weaker coupling, and it is convenient to consider the interactions stepwise in order of their decreasing magnitudes.

Spin coupling, being the strongest, is therefore considered first. The individual s vectors couple to give a resultant vector S, the numerical magnitude of which is $\sqrt{S(S+1)}\,h/2\pi$, where S is the resultant spin quantum number. When a number of s vectors couple they may do so in a variety of ways, depending on their inclinations to each other, and a number of possible values of S therefore result. Because the projection of s in a given direction is given by $m_s = \pm\frac{1}{2}$, the resultant spin quantum number is obtained by the algebraic summation of all the m_s values ($S = \sum m_s$) and can take integral or half integral values.

The next step is the coupling of the l vectors to give a resultant orbital angular momentum vector L. Again there are a variety of ways in which this may occur because of the different inclinations of the l's. Since the magnetic quantum number, m_l, takes account of the orientation of l, the value of L is obtained by the algebraic summation of all m_l values, $L = \sum m_l$. L can therefore take only integral values and the magnitude of the resultant L vector is given by $\sqrt{L(L+1)}\,h/2\pi$.

A great deal of confusion is likely if attempts are made to obtain the values of L or S graphically, particularly when more than two component l or s vectors are involved. This is a consequence of the fact, noted in Fig. 4, that the projection of such vectors in a given direction (e.g. the direction of the resultant) can never equal the value of the vector itself. It is for this reason that the simple algebraic summations have been used to obtain first the resultant quantum number and then, from this, the resultant angular momentum vector.

For each value of S there will, in general, be a number of possible values of L, each corresponding to a different energy. A particular pair of L and S values is known as a "term". Terms are conventionally represented by the capital letters, $S, P, D, F, G, H \ldots$ according as the value of L is 0, 1, 2, 3, 4, 5 …. The value of S is shown by the addition of a superscript, known as the "multiplicity" of the term, which is $2S+1$. Thus, for example, the terms ^6S, ^4D and ^2P signify ($S = \frac{5}{2}, L = 0$), ($S = \frac{3}{2}, L = 2$) and ($S = \frac{1}{2}, L = 1$).

It is not necessary in the present context to identify all the terms which may arise from a particular electronic configuration (this can be a complicated business), but clearly it is necessary to know which is energetically the lowest. The rules which allow this to be done were devised empirically by Hund from a study of atomic spectra. Their physical interpretation has been given above.

1. Of terms arising from a particular configuration, those with the maximum multiplicity (i.e. those corresponding to the most unpaired electrons) lie lowest.

2. If there is more than one term with maximum multiplicity, the lowest is that with the highest value of L.

These rules will also frequently indicate the order of the higher terms but contributions by other types of coupling for atoms in excited states may, in some cases, render them unreliable.

Finally, spin-orbit coupling occurs between the S and L vectors for each term giving a resultant, J, the S and L vectors precessing about the direction of J. The corresponding quantum number, J, can take the values $L+S$, $L+S-1, \ldots L-S$ (if $L>S$) or $S-L$ (if $S>L$) which may therefore be integral or half integral.

This is not, of course, pure $s_i \, l_i$ coupling since the spin and orbital vectors of the individual electrons cease to have exact significance when they have

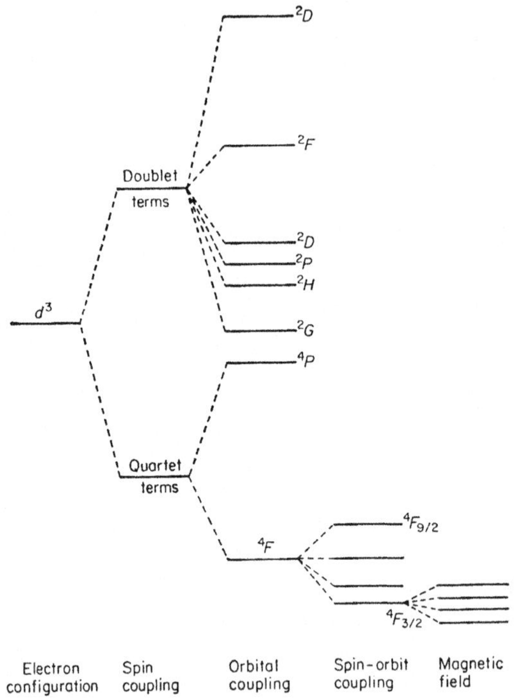

FIG. 6. The splitting of the free ion d^3 configuration under the influence of Russell-Saunders coupling and a magnetic field.

coupled to produce S and L. However, because of the fields associated with S and L, the coupling is still magnetic in origin and of similar form. In particular its strength is still dependent on the effective nuclear charge of the atom.

The collection of possible values of J for a given term is known as a "multiplet" and each J value is a "component" of that multiplet. Alternatively the term is said to split into "states". A particular state, or component, may be represented by adding the value of J as a subscript to the symbol for the term. The difference in energy between adjacent components is known as the "multiplet width". To decide whether the components of a multiplet are arranged energetically in order from the lowest to highest, or highest to lowest values of J, a third rule is used:

The component of a multiplet lying lowest is that with $J = L - S$ (when the multiplet is said to be "regular") if the electron shell is less than half full, and that with $J = L + S$ (when the multiplet is said to be "inverted") if the electron shell is more than half full.

It is evident that the number of components of a multiplet is either $2S + 1$ or $2L + 1$, depending on whether $L > S$ or $S > L$. The expression "the multiplicity" of a term is, however, conventionally taken to refer only to $2S + 1$ whether or not this is in fact the number of components of the multiplet.

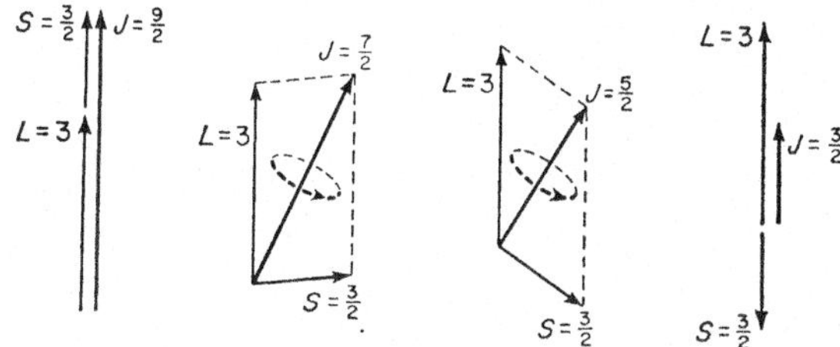

FIG. 7. Simple representation of the way in which different values of J arise.

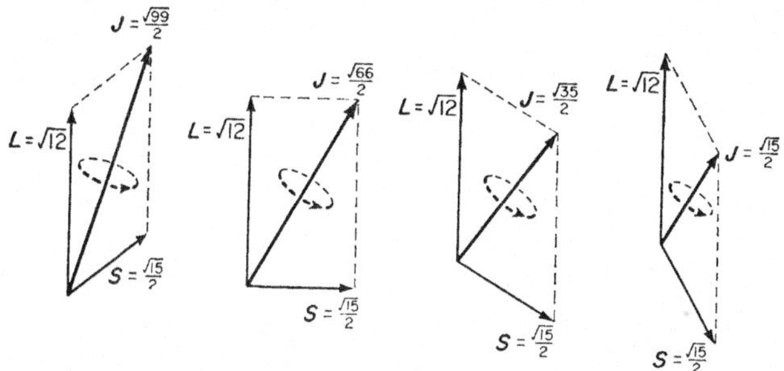

FIG. 8. A more accurate representation of the precession of L and S producing different values of the resultant J.

The operation of Russell-Saunders coupling will best be seen by taking a specific example, say the free ion d^3 configuration (Fig. 6). It is clear that only the three d electrons need to be considered since, for the filled shells of electrons, $\sum m_s = 0$ and $\sum m_l = 0$ so that they contribute nothing to the resultant angular momentum.

There are only two ways in which the spins of three electrons can couple. They may all be parallel, when $S = \frac{3}{2}$, or two may be "paired" with one remaining "unpaired", when $S = \frac{1}{2}$. Since these correspond to very different energies, the unperturbed d^3 configuration is said to be split by spin coupling into quartet and doublet levels. Hund's first rule indicates that the quartets lie lowest.

Orbital coupling splits these levels further. For d electrons $l = 2$ and m_l varies from $+2$ to -2 so that

$$m_l = \quad +2 \;\; +1 \;\;\; 0 \;\;\; -1 \; -2$$

$$\boxed{\;\uparrow\;|\uparrow\;|\uparrow\;|\;\;|\;\;\;}$$

For three unpaired electrons the maximum value of L consistent with the exclusion principle is 3. It follows therefore from Hund's second rule that a ^4F term lies lowest. The other terms produced are shown in Fig. 6 where it can be seen that the actual splitting of the doublet levels is evidence of the fact that Hund's rules do not always apply to excited states.

The multiplet structure of each term is produced next by the action of spin-orbit coupling. For the ^4F term, $S = \frac{3}{2}$, $L = 3$ and J can have values $\frac{9}{2}, \frac{7}{2}, \frac{5}{2}$ and $\frac{3}{2}$ as shown in Fig. 7.

This is a simple but naive representation in which the vectors are proportional in length to the corresponding quantum numbers. In Fig. 8 the vectors are drawn proportional to $\sqrt{S(S+1)}$ etc., illustrating that, even in the extremes of "parallel" and "antiparallel" alignment of S and L, these vectors still have components perpendicular to the direction of J.

Since the electron shell is less than half full the multiplet is regular and $^4F_{3/2}$ is its lowest component. The splitting of the ^4P and other terms can be obtained similarly.

Finally application of a magnetic field splits each component into $2J+1$ levels corresponding to the spacial quantization of J in the direction of the field. The projection of J is the total magnetic quantum number, M_J, the values of which range from $+J$ to $-J$. This is exactly analogous to the spacial quantization of l except that since J can have half integral as well as integral values so too can M_J. Diagrams similar to Fig. 6 can be drawn for all the d^x ions but it is often sufficient merely to know the ground term. Hund's rules readily indicate that these are

$$x = \;\; 0 \;\; 1 \;\; 2 \;\; 3 \;\; 4 \;\; 5 \;\; 6 \;\; 7 \;\; 8 \;\; 9 \;\; 10$$
$$\;^1S \; ^2D \; ^3F \; ^4F \; ^5D \; ^6S \; ^5D \; ^4F \; ^3F \; ^2D \; ^1S$$

Although there are obviously wide variations from element to element, the energy separation between adjacent terms in free ions is usually of the order of 10 000 cm^{-1}. Between different states, or components, of the same term the separation due to spin-orbit coupling is usually of the order 100–1 000 cm^{-1}. The splitting by the magnetic field is then of the order of a few wave numbers for the sort of field (say 10 000 gauss) usually encountered.

jj COUPLING

In this case the magnetic forces responsible for spin-orbit coupling are sufficient to overcome the interelectronic repulsions, and the situation is the reverse of that found when *LS* coupling applies. This time the spin and orbital angular momentum vectors of the individual electrons are coupled to give resultants j_i, j_k etc. where $j_i = l_i \pm s_i = l_i \pm \frac{1}{2}$. Because the coupling is strong the different j's will correspond to markedly different energies. A weaker coupling then occurs between the individual j vectors to give J for the whole atom. As before, the actual values of the angular momentum in units of $h/2\pi$ are $\sqrt{j(j+1)}$ and $\sqrt{J(J+1)}$ etc., and for filled shells the resultant is zero, as in *LS* coupling. Indeed the number of terms and the actual J values produced are the same on both schemes, though naturally their energies are quite different.

Pure *jj* coupling is rarely of interest, but frequently the actual coupling is found to be intermediate between *LS* and *jj*. This is to some extent true, for instance, in the second and third rows of the transition elements. In such cases, providing the coupling is predominantly *LS*, the spin-orbit coupling may be

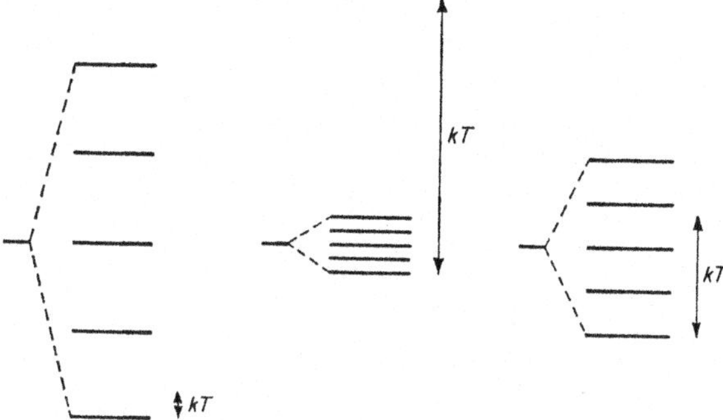

FIG. 9. Multiplet widths may be wide, narrow or comparable to kT.

assumed to act as a "perturbation" on the pure *LS* scheme. Where the contributions of the two schemes are comparable, as in the actinides, the treatment is far more complicated.

It has so far been shown how, on the basis of Russell-Saunders coupling, an electronic configuration gives rise to a number of possible situations in which the resultant angular momentum of the atom differs. These different situations, moreover, correspond to different energies. Obviously in any bulk sample of a substance a majority of the atoms will tend to be in the state of lowest energy, with the angular momentum and hence the magnetic moment appropriate to this. Some atoms, however, may possess sufficient energy for their electrons to produce a resultant angular momentum corresponding to a higher level, with a different magnetic moment. The proportion of atoms which are able to occupy higher levels will depend on the actual energies of these levels and on the thermal energy, kT, available to the atoms. In general, therefore, the magnetic properties of a substance will depend on the thermal population of different energy levels. Three distinct types of behaviour arise, depending on the relative magnitudes of kT and the multiplet widths. The multiplet widths may be large or small compared to kT, or they may be of comparable size (Fig. 9).

MULTIPLET WIDTHS LARGE COMPARED TO kT

When this occurs the L and S vectors are precessing rapidly about their resultant J and only a quite negligible proportion of the atoms possess sufficient energy to alter the coupling corresponding to the lowest energy, i.e. only the lowest component of the multiplet is populated to any significant extent, and in deducing the magnetic properties only this J value need be considered.

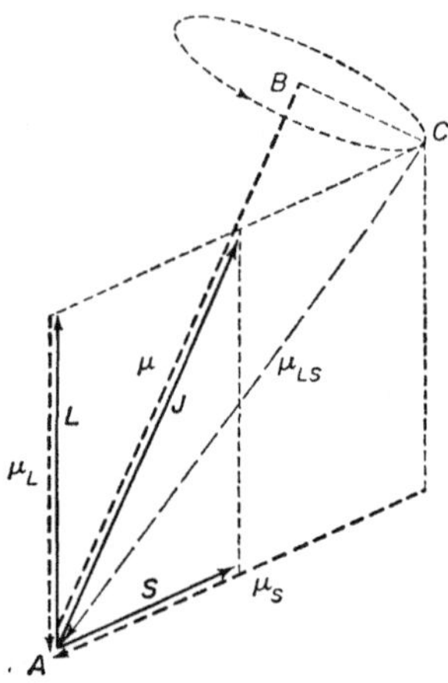

Fig. 10. The coupling of L and S and the associated magnetic moments. Because of the magnetic anomaly of spin the resultant magnetic moment, μ_{LS}, is not co-directional with the resultant angular momentum vector, J.

However, J is compounded of L and S and it has already been shown that spin and orbital angular momenta give rise to magnetic moment differently. In fact $\mu_L = \beta \sqrt{L(L+1)}$ and $\mu_S = 2\beta \sqrt{S(S+1)}$. As a consequence the resultant magnetic moment will not be co-directional with the resultant angular momentum. This is illustrated in Fig. 10 which shows both the angular momentum vectors (full lines) and the magnetic moment vectors (dashed lines). These are in exactly opposite directions and for convenience the scale is such that the μ_L vector is equal in length to the orbital angular momentum vector.

The resultant magnetic moment, μ_{LS}, is given by the vector AC but, because of the rapid precession of μ_L and μ_S about the direction of J, it may be taken as a good approximation that the component BC averages out to zero in any finite time and only its component μ in the direction of J need be considered.

The components of L and S in the direction of J are respectively L cos (LJ) and S cos (LJ).

Using the cosine rule

$$S^2 = J^2 + L^2 - 2LJ \cos (LJ)$$

and

$$L^2 = J^2 + S^2 - 2SJ \cos (SJ)$$

$$\therefore \quad L \cos (LJ) = \frac{J^2 + L^2 - S^2}{2J}$$

and

$$S \cos (SJ) = \frac{J^2 + S^2 - L^2}{2J}$$

The effective resultant moment, μ, is the sum of the moments associated with these component angular momenta. Therefore, converting angular momentum to magnetic moment and remembering the magnetic anomaly of spin:

$$\mu = \beta L \cos (LJ) + 2\beta S \cos (SJ)$$

$$= \left[\frac{J^2 + L^2 - S^2}{2J} + \frac{2J^2 + S^2 - L^2}{2J} \right] \beta$$

$$\therefore \quad \mu = \left[\tfrac{3}{2} + \frac{S^2 - L^2}{2J^2} \right] \beta J = g\beta J$$

Or, expressing L, S and J in terms of the quantum numbers L, S and J,

$$\mu = g\beta \sqrt{J(J+1)} \quad \text{e.m.u.} \tag{6}$$

where

$$g = \tfrac{3}{2} + \frac{S(S+1) - L(L+1)}{2J(J+1)} \tag{7}$$

In this case g is known as the "Lande splitting factor" since Lande derived equation (7) empirically to explain the spectroscopic splittings associated with the anomalous Zeeman effect.

As is shown by equations (3) and (6), g is the ratio of magnetic moment to angular momentum expressed in their respective quantum units. If the type of coupling differs from that under discussion then, of course, this ratio will not be given by the Lande formula, equation (7), and the values of g may vary considerably. In the Lande formula it can be seen that g may vary between 1 and 2 as J varies between L, $(S = 0)$ and S, $(L = 0)$. This is similar to the case of a single electron for which it has already been shown that $g_l = 1$ and $g_s = 2$.

Having obtained μ it is now possible to deduce χ_A, the magnetic susceptibility of the atom, by examining the energetics of the interaction between μ and an applied field, H. This may be done by considering the effect of H on a magnetic dipole, for which the magnetic moment is the product of the pole strength and the magnetic length, $\mu = mx$.

B

In a uniform field each pole, m, experiences a force mH (Fig. 11) producing a couple on the dipole. Therefore the moment of the couple is $mHx \sin \theta = \mu H \sin \theta$.

Therefore the work done by field in rotating the dipole through 90° (i.e. from a position perpendicular to H to one parallel to H) is $E = \mu H$.

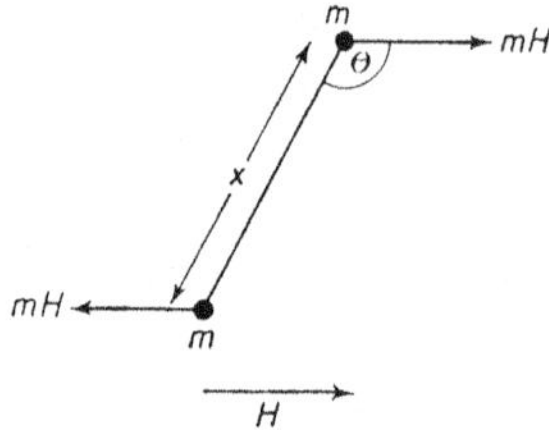

FIG. 11. A magnetic dipole in a field, H.

Incidentally, if the field is not uniform the forces on the two poles are unequal and, besides a torque on the dipole, there is a resultant force pulling the dipole into regions where H is stronger. This is the reason why paramagnetic substances are attracted into stronger fields and, conversely, why diamagnetic materials are repelled.

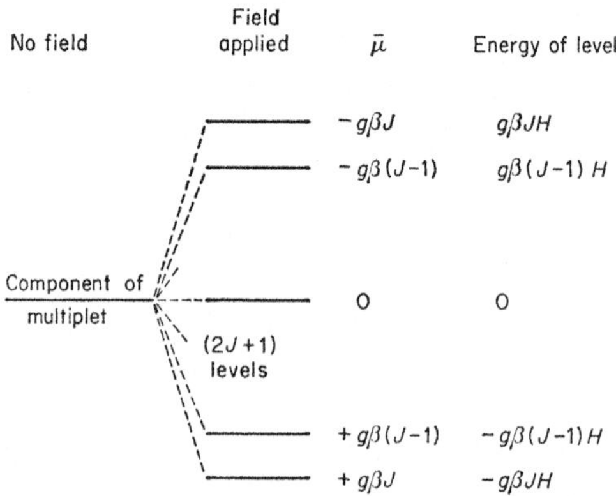

FIG. 12. The removal of the degeneracy of a component of a multiplet by the application of a magnetic field.

As was mentioned previously (p. 20), the projection, M_J, of J in the direction of H is quantized and can take values of $J, J-1, \ldots -J$. There are thus $2J+1$ values of M_J, each corresponding to a projection, $\bar{\mu}$, of μ. $\mu = g\beta \sqrt{J(J+1)}$ and $\bar{\mu}$ can therefore take values $-g\beta J, -g\beta(J-1), \ldots +g\beta J$, the signs of $\bar{\mu}$ being reversed with respect to the corresponding M_J since, as has already been noted, the magnetic moment is in exactly the opposite direction to the associated angular momentum vector.

The total spread of these levels is

$$2\mu H = 2g\beta J H$$

and it follows that the separation between adjacent levels is $g\beta H$.

It may be noted in passing that the Zeeman effect, which is the splitting of spectral lines on application of a magnetic field, is produced by this removal of the degeneracy of M_J. Thus transitions between particular components, whether of the same multiplet or not, will vary slightly in energy depending on the change occasioned in M_J ($\Delta M_J = 0, \pm 1$). The normal Zeeman effect occurs for singlet lines and the anomalous Zeeman effect for non-singlets. For a full account of this a standard work on spectroscopy should be consulted (see References, group 9).

In Chapter I, I was defined as the magnetic moment per unit volume. If we assume the material to consist entirely of monatomic molecules ($\chi_A = \chi_M$) and ignore all diamagnetic effects, then it follows that

$$I = \frac{\sum n\bar{\mu}}{V}$$

where n is here the number of atoms per gram mole with a particular projection of moment, $\bar{\mu}$, and V is the molar volume. But

$$\chi = \frac{I}{H\rho} = \frac{I}{H}\frac{V}{M} \text{ where } M \text{ is the molecular weight}$$

$$\therefore \quad \chi_A = \frac{IV}{H} = \frac{\sum n\bar{\mu}}{H} \tag{8}$$

In order to determine the population of each level the Boltzmann distribution law is applied. The ratio of the number of molecules in the different levels is then seen to be

$$\exp\left(-\frac{g\beta J H}{kT}\right), \ldots \exp(0), \ldots \exp\left(\frac{g\beta J H}{kT}\right)$$

Since $g\beta H \ll kT$ this reduces to:

$$\left(1 - \frac{g\beta J H}{kT}\right), \ldots (1-0), \ldots \left(1 + \frac{g\beta J H}{kT}\right)$$

But for 1 g mole of material the total number of atoms is Avogadro's number N and, since there are $2J+1$ levels, the actual numbers of atoms in each level are

$$\frac{N}{(2J+1)}\left[1 - \frac{g\beta J H}{kT}\right], \cdots \frac{N}{(2J+1)}, \cdots \frac{N}{(2J+1)}\left[1 + \frac{g\beta J H}{kT}\right]$$

and $n\bar{\mu}$ for each level is

$$\frac{N}{(2J+1)}\left[-g\beta J+\frac{g^2\beta^2 J^2 H}{kT}\right],\ldots 0,\ldots \frac{N}{(2J+1)}\left[g\beta J+\frac{g^2\beta^2 J^2 H}{kT}\right]$$

On summation of these values the first term in each pair of brackets will cancel, leaving

$$\sum n\bar{\mu}=\frac{N}{(2J+1)}\cdot\frac{g^2\beta^2 H}{kT}[J^2+(J-1)^2\ldots 0\ldots(J-1)^2+J^2]$$

$$=\frac{2N}{(2J+1)}\cdot\frac{g^2\beta^2 H}{kT}[J^2+(J-1)^2\ldots 0]$$

Now the sum, to the ith term, of a series of squares is

$$S_i=\frac{i(i+1)(2i+1)}{6}$$

$$\therefore\quad \sum n\bar{\mu}=\frac{2N}{(2J+1)}\cdot\frac{g^2\beta^2 H}{kT}\cdot\frac{J(J+1)(2J+1)}{6}$$

$$=\frac{Ng^2\beta^2 H}{3kT}\cdot J(J+1)$$

$\therefore$ from equation (8), $\chi_A=\dfrac{Ng^2\beta^2}{3kT}J(J+1)$ \hfill (9)

Combining equation (9) with equation (6)

$$\chi_A=\frac{N\mu^2}{3kT} \tag{10}$$

where μ is measured in e.m.u.

$$\text{or}\qquad \chi_A=\frac{N\beta^2\mu^2}{3kT} \tag{11}$$

where μ is measured in Bohr magnetons.

Equation (11) is the same as equation (5) in Chapter I, but with μ replaced by the "effective magnetic moment", μ_e. μ_e is defined by this equation and is "theoretical" or "experimental" according to whether χ_A is "theoretical" or "experimental". Irrespective of which way it is obtained it is identified with the permanent moment, μ, only when the Curie law is obeyed. Equation (10) is the formula of Langevin which he produced by integration over all space, assuming μ could take up any orientation. It is perhaps a rather surprising fact, known as the principle of spectroscopic stability, that the same result is obtained regardless of the type of quantization assumed. The more rigorous treatment of Van Vleck also yields essentially the same result, though a term $N\alpha$ is included which takes account of the small underlying diamagnetism and also the component of magnetic moment, perpendicular to J, which makes a temperature independent paramagnetic contribution to the susceptibility. This is the component BC in Fig. 10 which should strictly be ignored only when the multiplet width is infinitely great. However, under the condition that the multiplet widths are large, $N\alpha$ will certainly be very small and to a good approximation the Curie law will be obeyed, the susceptibility being independent of field and inversely proportional to temperature.

Multiplet Widths Small Compared to kT

The thermal energy available is now sufficient to populate all levels of the lowest lying multiplet to an effectively equal extent. Alternatively it can be said that the precession of L and S about J is so slow that on application of a magnetic field it is the resultant magnetic moment (AC in Fig. 10), rather than its component in the direction of J, which is oriented, and hence quantized, in the direction of the field. This is effectively the same as L and S interacting separately with the field. If this happens then $\mu_L = \beta \sqrt{L(L+1)}$ and $\mu_S = 2\beta \sqrt{S(S+1)}$ and, by a procedure exactly analogous to that used for wide multiplets, orbital and spin susceptibilities can be worked out and the total atomic susceptibility obtained by their addition.

$$\chi_A = \frac{N\beta^2}{3kT}[L(L+1)+4S(S+1)]$$

whence

$$\mu = \beta\sqrt{L(L+1)+4S(S+1)} \tag{12}$$

Once again a small correction should strictly be added for the underlying diamagnetism but on this occasion there is no temperature independent paramagnetic term. Again the Curie law is obeyed.

Multiplet Widths Comparable to kT

With multiplet widths of intermediate size, the situation is a mixture of the two extremes already dealt with. All, or at least several, components of the lowest lying multiplet will be significantly populated but to differing extents. The average susceptibility of a sample of the material is the result of contributions from each component of the multiplet "weighted" according to its population.

The expression for the atomic susceptibility (equal to the molar susceptibility on the assumption of monatomic molecules) due to complete population of a single component has already been given in equation (9). In the present case the contribution to the susceptibility, of each level, is given by substituting N_J for N, where N_J is the number of molecules per gram molecule with that particular value of J.

Neglecting $N\alpha$ terms* we must therefore sum a set of expressions of the form

$$\frac{N_J g^2 \beta^2 J(J+1)}{3kT}$$

Now the number of molecules in a *single* level, of energy E above the ground, is proportional to $\exp(-E/kT)$. However, each component has $2J+1$ possible orientations, each of which counts as a separate state, and so each level must be weighted accordingly

$$\therefore \quad N_J \propto (2J+1) \exp\left(-\frac{E}{kT}\right) = C(2J+1)\exp\left(-\frac{E}{kT}\right)$$

where C is the proportionality constant.

Summation over all values of J gives

$$\chi_A = C \sum \frac{g^2\beta^2 J(J+1)}{3kT}(2J+1)\exp\left(-\frac{E}{kT}\right) \tag{13}$$

but N_J is defined by $N = \sum N_J$

$$\therefore \quad N = C\sum(2J+1)\exp\left(-\frac{E}{kT}\right)$$

$$\therefore \quad C = \frac{N}{\sum(2J+1)\exp\left(-\frac{E}{kT}\right)}$$

$\therefore$ equation (13) becomes

$$\chi_A = \frac{\dfrac{N}{3kT}\sum g^2\beta^2 J(J+1)(2J+1)\exp\left(-\dfrac{E}{kT}\right)}{\sum(2J+1)\exp\left(-\dfrac{E}{kT}\right)} \tag{14}$$

The temperature dependence of the susceptibility is this time far removed from the Curie law.

The results for the different multiplet widths, on the assumption of **LS** coupling, are summarized in Table IV.

In obtaining these results certain assumptions have been made. In the first place, as was mentioned, the underlying diamagnetism and, where appropriate, the temperature independent paramagnetic term have been ignored. This is usually not too serious except for substances with low moments at room

---

* Since we shall not be concerned in treating this case quantitatively, $N\alpha$ may be omitted for simplicity, but it should be realized that on occasions this term may be of great importance. This happens with Eu^{3+} and Sm^{3+} which will be mentioned later.

temperature and above. More serious is the tacit assumption that each atom may be dealt with as if independent of all other atoms. This is rather extreme and it might be thought likely to apply only to monatomic gases. Unfortunately the common monatomic gases are all diamagnetic but the point is well illustrated by the case of potassium. Here the ground state is an S term, there is no multiplet structure, and $J = S = \frac{1}{2}$. The expected moment is therefore $\sqrt{3}$ B.M. Gerlach has shown that the vapour obeys the Curie law with a moment close to this value though, due to obvious experimental difficulties, the accuracy is not claimed to be very high. This behaviour is in marked contrast to that for solid potassium and other alkali metals where interatomic interactions are so great that the paramagnetism almost disappears.

The most valuable confirmation of the preceding theory is frequently thought to be provided by the lanthanide elements and their compounds.

TABLE IV

Multiplet width compared to kT	μ in Bohr magnetons	Temperature dependence of χ_A
Large	$g\,\sqrt{J(J+1)}$	Curie law obeyed
Small	$\sqrt{L(L+1)+4S(S+1)}$	Curie law obeyed
Comparable	Complicated function of J and T	Curie law *not* obeyed

In them the unpaired $4f$ electrons responsible for the paramagnetism are quite effectively screened from environmental effects by overlying s and p electrons. Even in the oxides and the elements themselves, where the proximity of the atoms or ions is closest, there is little evidence of any significant interaction between the electrons of neighbouring ions. It may thus be thought that each ion would be able to orient itself freely in a magnetic field like an ion in the gaseous state. With the exceptions of europium and samarium, the multiplet widths of the lanthanides are wide compared to kT ($kT \simeq 200$ cm^{-1} at room temperature), consequently equation (6) should apply, with the Curie law being obeyed. In fact, however, the effect of the electric charges associated with the surrounding ligands is to lift the degeneracy of individual states by an amount of the order of 100 cm^{-1}. Though this is small compared to the separations between individual states, produced by spin-orbit coupling, it is not negligible compared to kT. Contrary to the prediction of equation (6) the moments of these ions should therefore be temperature dependent. This is actually the case since most of the ions obey the Curie-Weiss law but with appreciable values of θ. The apparent applicability of equation (6), which is illustrated in Table V, is due to the fortuitous coincidence of the predicted values with measurements at room temperature. When measurements are made at lower temperatures this coincidence is not found. An exact evaluation of the effect of surrounding ligands is extremely difficult, due to the complexity not only of the f orbitals but also of the electric fields involved in lanthanide compounds.

Most measurements have been on the sulphates and oxides. Narrower multiplet widths occur in the cases of europium and samarium and necessitate the use of equation (14) with values of $N\alpha$ included. In this way Van Vleck has accounted quite satisfactorily for the rather complicated temperature dependence of the moments of the trivalent ions of these elements.

TABLE V

No. of f electrons	M^{3+}	Ground state	g	μ_e (B.M.) Calc. from (9)	μ_e (B.M.) Expt (room)
0	La	$^1S_0$	—	0	Diamagnetic
1	Ce	$^2F_{5/2}$	6/7	2·54	2·3–2·5
2	Pr	$^3H_4$	4/5	3·58	3·4–3·6
3	Nd	$^4I_{9/2}$	8/11	3·62	3·5–3·6
4	Il	$^5I_4$	3/5	2·68	—
5	Sm	$^6H_{5/2}$	2/7	0·84	1·5–1·6
6	Eu	$^7F_0$	—	0	3·4–3·6
7	Gd	$^8S_{7/2}$	2	7·94	7·8–8·0
8	Tb	$^7F_6$	3/2	9·72	9·4–9·6
9	Dy	$^6H_{15/2}$	4/3	10·63	10·4–10·5
10	Ho	$^5I_8$	5/4	10·60	10·3–10·5
11	Er	$^4I_{15/2}$	6/5	9·57	9·4–9·6
12	Tm	$^3H_6$	7/6	7·63	7·1–7·4
13	Yb	$^2S_{7/2}$	8/7	4·50	4·4–4·9
14	Lu	$^1F_0$	—	0	Diamagnetic

Although in the past the chief importance of magnetic measurements on lanthanide ions has probably been to verify magnetic theory (albeit fortuitously), it may be hoped that future developments will allow sensitive measurements of temperature dependence to yield information relating to the stereochemistry of their compounds.

Another example of inner orbitals being filled occurs with the transition elements. However, the simple ions of these elements contain no electrons of principal quantum number higher than that of the d electrons responsible for the paramagnetism. Screening, of the kind found in the lanthanides, is thus impossible and appreciable modification of the theory is needed because of the marked influence of environment. It is because of this dependence of magnetic properties on environment that magnetic measurements are so useful in studying transition elements and their compounds. This will be dealt with more fully in Chapter III.

Although the matter will not be pursued, it is of interest to note that a similar approach to diatomic gases can be used to explain the magnetic properties of oxygen and nitric oxide. The oxygen molecule, with two unpaired electrons, has a triplet ground state with multiplet widths very small compared to kT and zero orbital angular momentum. Its theoretical molar susceptibility is therefore

$$\chi_M = \frac{N\beta^2}{3kT}[4S(S+1)] = \frac{8N\beta^2}{3kT} = 3\,390 \times 10^{-6} \text{ at } 20°\text{C}$$

It is also expected to obey the Curie law. Experimental results agree closely with this.

Available information on sulphur vapour suggests, as would be expected, that the S_2 molecule is closely similar to O_2.

Nitric oxide, with a single unpaired electron, has a doublet ground state the width of which is comparable to kT at room temperature, and shows rather complicated magnetic behaviour. Again, however, Van Vleck, using an approach analogous to that for europium and samarium, has satisfactorily accounted for this.*

* For references for this chapter, see groups 1, 9 and 14 on pp. 111 and 112.

B*

III. TRANSITION METAL COMPLEXES

The chemistry of the transition elements provides the most interesting and fruitful field for magnetic investigation. As has been remarked, this is due to the fact that the influences of neighbouring groups on the d electrons of the metal ions are sufficiently strong to affect significantly their magnetic properties. Providing the mechanisms of these interactions are understood, the study of magnetic properties can yield valuable information about the bonding and structures of transition metal compounds. There are, in the main, three methods of accounting for the influences of surrounding molecules or ions on the d electrons of transition metal ions. The first is the valence bond theory of Pauling, which is primarily concerned with the bonding involved, and considers the stereochemistry to result from the orbitals used and the type of hybridization which occurs. The bonding was initially assumed to be either wholly covalent or wholly ionic, though this was subsequently modified to "inner orbital" and "outer orbital", referring to the orbitals on the metal used to accommodate the covalently bonding electrons from the attached groups. Although this approach is now mainly of historical interest it stimulated much of the work responsible for our present understanding of co-ordination chemistry, and is deeply woven into the fabric of inorganic terminology.

The second approach is crystal field theory due to Bethe and Van Vleck, which assumes the bonds between ligands and metal to be purely ionic. It is primarily concerned with the electrostatic effect of the ligands on the non-bonding electrons of the metal ion, and can be considered as a special case of the third approach, ligand field theory. In this the possibility of covalency is included by using the molecular orbital theory of Mulliken. It is naturally somewhat more complicated and, since the results of crystal field theory are nearly all equally valid in ligand field theory (for this reason the names are sometimes used rather loosely), attention will be restricted to the former.

Throughout this chapter the condition of magnetic dilution will be assumed.

VALENCE BOND THEORY

Pauling was the first to apply this theory to co-ordination chemistry and thereby to show clearly the connexion between the structure of a transition metal compound and its magnetic moment.

The formation of a co-ordination compound is assumed to occur in the following, hypothetical, manner. The metal first loses electrons, the number of which is the oxidation state of the cation so formed. Hybridization of appropriate orbitals on the ion then occurs so as to facilitate their occupation by electron pairs from the co-ordinated groups. The disposition of these groups and the hybridized orbitals which they use, define the stereochemistry of the molecule. The non-bonding electrons of the metal ion, which are assumed to be of secondary importance, are then arranged in accordance with Hund's rule that the most stable arrangement is the one with the maximum number of unpaired electrons, in whatever orbitals remain. The types of hybridization occurring most commonly in first row transition metal compounds are:

sp^3 or d^3s	4 co-ordinate, tetrahedral
dsp^2	4 co-ordinate, square planar
dsp^3	5 co-ordinate, square pyramidal or trigonal bipyramidal
sp^3d^2	6 co-ordinate, octahedral
d^2sp^3	6 co-ordinate, octahedral

The two octahedral forms are known as "ionic" or "outer orbital" and "covalent" or "inner orbital" complexes, since they make uses of inner and outer d orbitals respectively ($3d$ and $4d$ for first row transition elements). Since the number of d orbitals utilized for bonding is dependent on the stereochemistry and bond type of the complex, it follows that the arrangement of the non-bonding electrons in the remaining d orbitals must also depend on these factors. Also, of course, the actual number of non-bonding electrons is dependent on the oxidation state of the metal ion. These points are well illustrated by manganese in its valencies of 2 and 3:

	3d					4s	4p			4d					n
Mn^{2+}	↑	↑	↑	↑	↑										5
tet. sp^3	↑	↑	↑	↑	↑	xx	xx	xx	xx						5
oct. sp^3d^2	↑	↑	↑	↑	↑	xx	xx	xx	xx	xx	xx				5
oct. d^2sp^3	↑↓	↑↓	↑	xx	xx	xx	xx	xx	xx						1
Mn^{3+}	↑	↑	↑	↑											4
oct. sp^3d^2	↑	↑	↑	↑		xx	xx	xx	xx	xx	xx				4
oct. d^2sp^3	↑↓	↑	↑	xx	xx	xx	xx	xx	xx						2

where n is the number of unpaired d electrons, tet. is tetrahedral, oct. is octahedral, $\uparrow$ or $\downarrow$ represent a non-bonding metal electron and x a bonding ligand electron.

Since the electrons in an incompletely filled shell give rise to a finite result-ant angular momentum, the magnetic moment of a complex will depend, in an as yet unspecified way, on the actual number of unpaired electrons. Although the correspondence between n and the stereochemistry, bond type and valency is not always unique, it is obvious that measurement of magnetic moment, particularly in conjunction with other chemical evidence, will be extremely helpful in distinguishing between the various possibilities. Table VI gives the

TABLE VI

Ions	Ground term	μ_e (B.M.)			
		$[L(L+1)+4S(S+1)]^{\frac{1}{2}}$	$g[J(J+1)]^{\frac{1}{2}}$	$[4S(S+1)]^{\frac{1}{2}}$	Expt
Ti^{3+}	$^2D_{3/2}$	3·00	1·55	1·73	1·7–1·8
Ti^{2+}, V^{3+}	$^3F_2$	4·47	1·63	2·83	2·7–2·9
V^{2+}, Cr^{3+}	$^4F_{3/2}$	5·20	0·70	3·87	3·7–3·9
Cr^{2+}, Mn^{3+}	$^5D_0$	5·48	0	4·90	4·8–4·9
Mn^{2+}, Fe^{3+}	$^6S_{5/2}$	5·92	5·92	5·92	5·7–6·0
Fe^{2+}, Co^{3+}	$^5D_4$	5·48	6·71	4·90	5·0–5·6
Co^{2+}	$^4F_{9/2}$	5·20	6·63	3·87	4·3–5·2
Ni^{2+}	$^3F_4$	4·47	5·59	2·83	2·9–3·5
Cu^{2+}	$^2D_{5/2}$	3·00	3·55	1·73	1·8–2·1

values of μ_e calculated for the d^x configurations on the basis of small multiplet widths ($\mu_e = \sqrt{L(L+1)+4S(S+1)}$ B.M.), large multiplet widths ($\mu_e = g\sqrt{J(J+1)}$ B.M.) and by means of what is, for obvious reasons, known as the "spin-only" formula:

$$\mu_e = \sqrt{4S(S+1)} \text{ B.M.}$$

The assumption is made (and is reasonable for, say, heavily hydrated salts or their solutions) that these expressions can be evaluated by using the values of L, S and J of the free ion ground term. For comparison ranges of values, found experimentally for compounds of the appropriate di- and tri-valent first row transition element ions, are also included.

None of the theoretical formulae fits the experimental data perfectly but much the best is the "spin-only" expression which provides a reasonable fit, except for the later members of the series. The total spin quantum number is related to the number of unpaired electrons, n, by $S = n/2$. This provides an alternative way of expressing the "spin-only" formula:

$$\mu_{\text{s.o.}} = \sqrt{n(n+2)} \text{ B.M.} \tag{1}$$

The assumption is frequently made that, for first row transition elements, the multiplet widths are small compared to kT, whence the "spin-only" formula results providing $L = 0$. In cases where the moment exceeds $\mu_{s.o.}$ it is presumed that this is due to the presence of a finite value of L, i.e. of an

TABLE VII. *Magnetic moments of some complexes of first row transition metals*

Con-figuration	Example	Stereo-chemistry	Hybrid orbitals	n	μ_e (B.M.) Spin-only	Expt
d^1	K_3TiF_6	oct.	d^2sp^3	1	1·73	1·70
d^2	$V(acac)_3$	oct.	d^2sp^3	2	2·83	2·80
d^3	$[Cr(NH_3)_6]Br_3$	oct.	d^2sp^3	3	3·88	3·77
d^4	$CrClO_46H_2O$	oct.	sp^3d^2	4	4·90	4·97
	$[Cr(dipy)_3]Br_24H_2O$	oct.	d^2sp^3	2	2·83	3·27
d^5	$[Mn(py)_6]Br_2$	oct.	sp^3d^2	5	5·92	6·00
	$K_3[Mn(CN)_6]3H_2O$	oct.	d^2sp^3	1	1·73	2·18
	$(Et_4N)_2[MnCl_4]$	tet.	sp^3	5	5·92	5·94
	$Na_3[FeF_6]$	oct.	sp^3d^2	5	5·92	5·85
	$K_3[Fe(CN)_6]$	oct.	d^2sp^3	1	1·73	2·25
d^6	$[Fe(NH_3)_6]Cl_2$	oct.	sp^3d^2	4	4·90	5·45
	$K_4[Fe(CN)_6]$	oct.	d^2sp^3	0	0	0
	$[Fe(Ph_3P)_2]I_2$	tet.	sp^3	4	4·90	5·10
	Fe(II)phthalocyanine	planar	dsp^2	2	2·83	3·96
	$Na_3[CoF_6]$	oct.	sp^3d^2	4	4·90	5·39
	$[Co(NH_3)_6]Cl_3$	oct.	d^2sp^3	0	0	0
d^7	$[Co(en)_6]SO_4$	oct.	sp^3d^2	3	3·88	4·56
	$K_2Ba[Co(NO_2)_6]$	oct.	d^2sp^3	1	1·73	1·88
	$Co py_2Cl_2$ violet	oct.	sp^3d^2	3	3·88	5·15
	$Co py_2Cl_2$ blue	tet.	sp^3	3	3·88	4·42
	Co(II)phthalocyanine	planar	dsp^2	1	1·73	2·73
d^8	$[Ni(NH_3)_6]Cl_2$	oct.	sp^3d^2	2	2·83	3·32
	$[Ni(diars)_3](ClO_4)_2$	oct.	d^2sp^3	0	0	0
	$(Et_4N)_2[NiCl_4]$	tet.	sp^3	2	2·83	3·89
	$Ni(Et_3P)_2Cl_2$	planar	dsp^2	0	0	0
d^9	$[Cu(phen)_3](ClO_4)_2$	oct.	sp^3d^2	1	1·73	1·96
	$Cu(DMG) Cl_2$	planar	dsp^2	1	1·73	1·85

Though d^2sp^3 hybridization has been ascribed to the octahedral configurations d^1, d^2 and d^3, there is actually no way of distinguishing this from sp^3d^2.

"orbital contribution". However, the assumption of narrow multiplet widths is by no means always justified but, even with the opposite assumption of large multiplet widths, $\mu_{s.o.}$ is still obtained if the orbital contribution is zero since then $J = S$ and $g = 2$. For the d^5 configuration the calculated moment is the same in all cases simply because, from Hund's rules, $L = 0$.

If at this stage the rather surprising disappearance of the orbital contribution is simply accepted along with the applicability of equation (1), then a straightforward procedure emerges.

The magnetic moment is measured and n is deduced. This is then related to the possible arrangements as was outlined in the case of manganese. Table VII shows the relationship between μ_e and stereochemistry in a number of cases.

Table VIII summarizes the number of unpaired electrons expected for different stereochemistries for configurations d^1 to d^9. This must not be taken to imply that all these stereochemistries will actually occur for each configuration.

This method provides no means of distinguishing different stereochemistries or bond types up to, and including, the d^3 configuration but is of considerable use with the later ions.

It is usually possible to infer the valency of the metal ion by purely chemical

TABLE VIII. *Values of* n *for different stereochemistries and configurations*

No. of d electrons:	1	2	3	4	5	6	7	8	9
Octahedral, outer orbital sp^3d^2	1	2	3	4	5	4	3	2	1
Octahedral, inner orbital d^2sp^3	1	2	3	2	1	0	1	0	1
Tetrahedral sp^3	1	2	3	4	5	4	3	2	1
Square planar dsp^2	1	2	3	4	3	2	1	0	1

means but this is not always the case; there is always some doubt, for instance, whenever the NO group is present in a complex. The nitroprussides, $M_2[Fe(CN)_5NO]$, were originally thought to contain iron in the trivalent state with neutral NO. The compounds are, however, diamagnetic and the iron must be therefore formally inner orbital octahedral ferrous, with co-ordinated nitrosonium ion, NO^+. The composition of the brown ring, obtained when testing for nitrates, has presented similar problems. Aqueous solutions are readily obtained by the action of nitric oxide on ferrous solutions. The moments are about 3·9 B.M., suggesting a d^7 configuration with three unpaired electrons. This leads to the formulation $[Fe(H_2O)_5NO]^{2+}$, containing Fe^+ and NO^+ ions and in keeping with infrared evidence.

Even where the experimentally determined moments differ markedly from the spin-only values, the known results can be used as a standard with which compounds of unknown valency, stereochemistry or bond type may be compared. The most striking example of this is the cobaltous ion. Both tetrahedral and outer orbital octahedral complexes of this ion are expected to have three unpaired electrons. The experimental moments in fact seem to lie in two fairly distinct ranges: 4·4–4·8 and 4·8–5·2 B.M. respectively. Though it is not at this stage obvious why these ranges should be distinct, nor indeed why they should be so far in excess of the spin-only value of 3·88 B.M., their usefulness is evident.

For the Ni(II) ion the situation is reversed in that it is the tetrahedral complexes which have the higher moments. The two ranges are approximately 2·9–3·3 B.M. for octahedral and 3·2–4·0 B.M. for tetrahedral complexes. Appreciable overlapping occurs and assignment of stereochemistry purely on the basis of the moment is impossible if this is around 3·2 B.M. The complex, Ni(acetylacetonate)$_2$ for which $\mu_{300} = 3·20$ B.M., has frequently been cited as an example of tetrahedral nickel, whereas an X-ray structural determination has in fact shown it to be trimeric and octahedral.

A good deal of controversy has centred around certain compounds of Ni(II) which, in solution, have moments appreciably less than the spin-only value of 2·83 B.M. It appears that this is due to equilibria between diamagnetic square planar and paramagnetic tetrahedral or, more probably, octahedral forms. This is supported by the fact that the observed moments in such cases are critically dependent on solvent, concentration and temperature. This will be further considered in Chapter V.

While the valence bond approach, as outlined above, has proved to be extremely useful, it has several inherent limitations because of which its use is now restricted to a fairly elementary level.

1. There is no indication as to why, in some cases, moments are appreciably in excess of the spin-only values while in others, particularly in the first half of the transition series, they are somewhat lower. Indeed there is no reason why the spin-only values should be expected to apply at all.

2. With the exception of Co(II) and, to a lesser extent, Ni(II), the distinction between tetrahedral and outer orbital octahedral complexes is usually poor. The most satisfactory method of distinguishing the two is to examine the temperature dependence of the moment. However, there is nothing in the preceding account to suggest that temperature dependence should occur at all, and to account satisfactorily for it recourse must be made to more recent theories.

3. For d^7, d^8 and d^9 configurations, inner orbital octahedral complexes require promotion of electrons to higher orbitals. The fact that these orbitals are unspecified presents the difficulty that, whereas p or d orbitals would allow electrons to remain unpaired, an s orbital would necessitate pairing in the d^8 and d^9 cases. In Table VIII it is assumed, in accord with what experimental evidence there is, that promotion is to the $5s$ orbital (leading to $n = 0$ and 1 respectively) rather than to the $5p$ or $4s$ (leading to $n = 2$ and 3 respectively).

4. In transition metal complexes spectral and magnetic properties, both being dependent on the arrangement of the d electrons, must be interrelated. Valence bond theory provides no indication of how they may be correlated.

CRYSTAL FIELD THEORY

When a metal ion is surrounded by co-ordinated groups, or ligands, it must experience an electrostatic field set up by these groups. Even if the ligands are not negatively charged, they will be polarized by the positive charge of the cation and must therefore give the appearance of being charged. This electrostatic field will obviously exert some effect on the electrons of the cation—a possibility which is ignored in the valence bond approach.

The covalency of the molecule may take the form simply of electron donation from the ligands to the cation or, in addition, bonding or "back donation" from the cation to the ligands. Rather surprisingly, however, considerable progress can be made if the possibility of covalency is completely ignored, and the ligands, or at any rate the actual donor atoms, are considered as point negative charges. This purely electrostatic approach is known as "crystal field" theory because it was originally developed for ions in a crystal lattice. Since the major effects are due to nearest neighbours, it is evidently immaterial whether these are part of an extended lattice or are the donor atoms of ligands associated with a metal ion in a separate entity.

The most symmetrical electrostatic fields are spherical but these do not occur in complexes. Cubic fields are in practice the most symmetrical and occur in octahedral and tetrahedral complexes. These are simpler than the less symmetrical cases and will be dealt with first. Perfectly cubic fields are, of course, somewhat rare if for no other reason than the distorting effects of atoms or molecules outside the immediate co-ordination sphere. However, these distortions are frequently small enough to be ignored.

Octahedral Complexes

It is convenient to imagine an octahedral complex as being formed by six ligands, or donor atoms, approaching the central cation along its x, y and z axes. The relationship between these axes and the five d orbitals is shown in Fig. 13. These diagrams are rather inaccurate, but easily drawn, boundary surfaces which enclose the bulk of the electron density of the orbitals. In the free ion the orbitals are energetically identical or "degenerate". As the ligands approach, however, any electrons in the orbitals are repelled, i.e. the energy of the orbitals is increased. In addition their degeneracy is removed (Fig. 14), since the $d_{x^2-y^2}$ and d_{z^2} orbitals lie along the x, y and z axes and are thus destabilized more than the d_{xy}, d_{xz} and d_{yz} orbitals which lie between the axes. These two sets of orbitals are referred to as the d_γ or e_g set and the d_ε or t_{2g} set respectively. The difference in energy between them is Δ_0 or $10\,Dq$ which is a measure of the intensity of the electrostatic field acting on the metal ion. The more usual, though less informative, representation of the splitting of the metal d orbitals is shown in Fig. 15. It is this removal of the degeneracy of the d orbitals which is responsible for the quenching of the orbital contribution and the consequent applicability of the spin-only formula for the

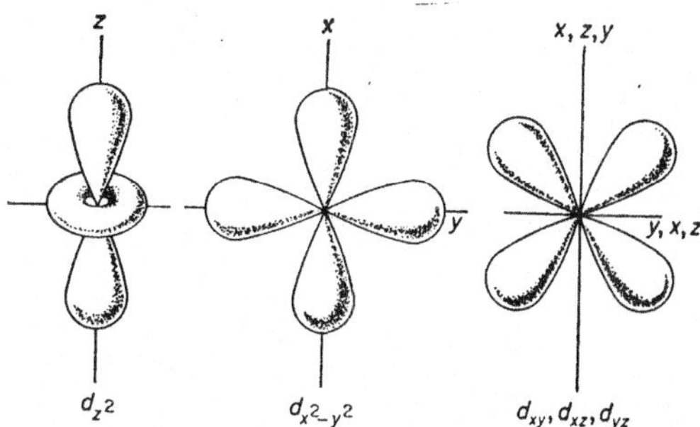

FIG. 13. Diagrammatic representations of d orbitals.

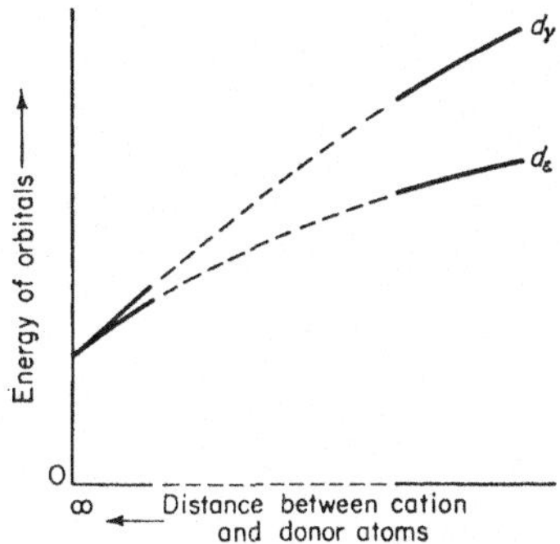

FIG. 14. The removal of the degeneracy of a set of d orbitals as an octahedral complex is formed.

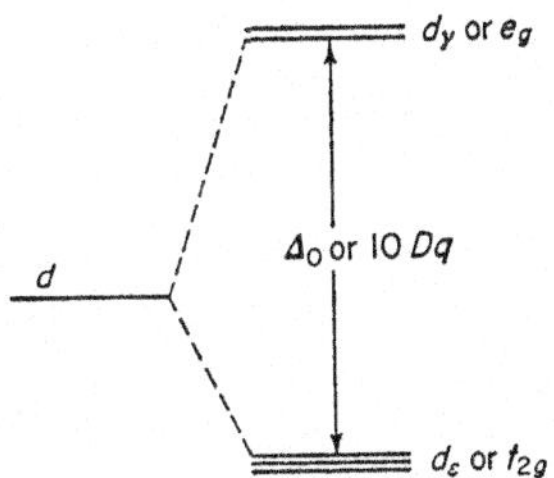

FIG. 15. The conventional representation of the splitting of d orbitals into doublet and triplet sets in an octahedral complex.

magnetic moment. This will be clearer if we digress slightly to see how orbital angular momentum is developed in free ion d orbitals.

It has previously been shown that orbital angular momentum could be thought crudely to arise from rotation of the electron about the nucleus, as in the Bohr model of the atom. This simple picture is complicated in the wave mechanical model of the atom where the orbital angular momentum may be pictorially associated with the interchange, or transformation, of one orbital with another by rotation about an appropriate axis. For this to be possible the orbitals must be degenerate and of the same shape and must not both contain electrons of the same spin. If these conditions are satisfied, then an electron in one of the orbitals will be able to rotate about the axis, being effectively an electric current producing a magnetic field in accord with the simple picture used before. The d_{xy}, d_{xz} and d_{yz} orbitals can transform into each other by 90° rotations about the relevant axes while, even more effectively, 45° rotations of the d_{xy} orbital about the z axis transform it into the $d_{x^2-y^2}$ orbital. For a single electron in a free ion all these rotations are possible and the full orbital angular momentum is developed. For the d^5 and d^{10}

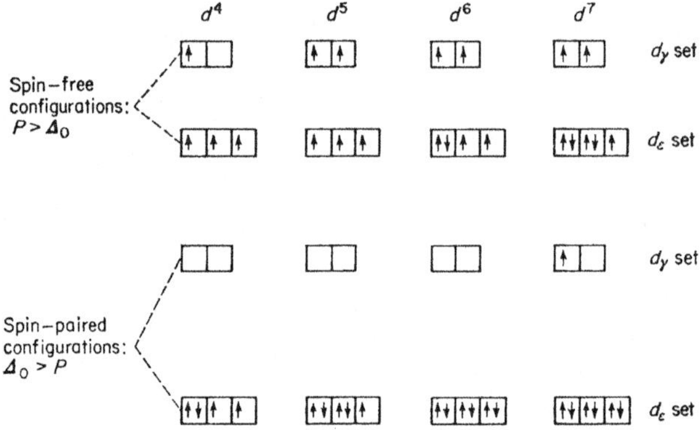

FIG. 16. The production of spin-free and spin-paired configurations in octahedral complexes of transition elements.

configurations, $L = 0$ and no orbital contribution to the magnetic moment is expected. These are the cases in which every orbital contains electrons of the same spin, thus preventing the development of any orbital angular momentum.

The effect of the imposition of an octahedral field on to the free ion can now be seen. In the first place the interchange of the d_{xy} and $d_{x^2-y^2}$ orbitals, which was an important source of orbital contribution in the free ion, is no longer possible. Secondly, although interchange within the d_ε set is still energetically feasible, the condition that orbitals must not contain electrons of the same spin is now more limiting. In the free ion this precluded orbital contribution in only the two configurations d^5 and d^{10}, whereas now the orbital contribution is quenched in all cases involving $d_\varepsilon{}^3$ or $d_\varepsilon{}^6$ configurations irrespective of the number of electrons in the d_γ set. It can be seen from Fig. 13 that the $d_{x^2-y^2}$ and d_{z^2} orbitals, being of different shapes, cannot be interchanged and hence have no orbital angular momentum associated with them. It is for this reason that the d_γ set is known as a "non-magnetic" doublet. This expression implies nothing about the spin angular momentum of electrons in the d_γ set, which remains unaffected. Electronic configurations involving $d_\varepsilon^{\,1}$, $d_\varepsilon^{\,2}$, $d_\varepsilon^{\,4}$ or $d_\varepsilon^{\,5}$ should still give rise to some orbital contribution leading to moments in excess of the spin-only values.

For d^1, d^2 and d^3 configurations the electrons enter the more stable d_ε set while, for d^8 and d^9, the unpaired electrons must be in the higher d_γ set. However, for each of the intervening configurations d^4, d^5, d^6 and d^7, there are two distinct possibilities, shown in Fig. 16. Which of these actually occurs depends on the relative magnitudes of two opposing effects. First the electrostatic field, represented by Δ_0, tends to prevent electrons entering the d_γ orbitals. Secondly the exchange energy (P) associated with Hund's first rule, which is the energy required to force two electrons to pair in the same orbital, tends to prevent electrons pairing in the d_ε orbitals. For first row d^4 to d^7 ions, P varies roughly between 20 000 and 30 000 cm^{-1}, while Δ_0 varies from 5 000 to over 30 000 cm^{-1}. With such an overlap it is obvious that in different circumstances either of these quantities may be greater than the other. Thus, if $\Delta_0 > P$, electrons will be forced into the lower orbitals giving "spin-paired" or "low-spin" compounds. If, on the other hand, $P > \Delta_0$, then Hund's first rule still operates and electrons spread out over the complete set of d orbitals giving "spin-free" or "high-spin" compounds.

To a certain extent valence bond theory and this simple crystal field theory may be regarded as complementary to each other. The bonding d orbitals in the inner orbital, octahedral complexes of valence bond theory are just those which, in crystal field theory, are considered to be too high for occupation by non-bonding electrons. Spin-paired or low-spin complexes are equivalent to inner orbital complexes and spin-free or high-spin correspond to outer orbital. Discrepancies exist, however, in the d^7, d^8 and d^9 configurations. Whereas the octahedral, inner orbital complexes of valence bond theory require electron promotion to a higher s orbital, no such promotion is needed on the basis of crystal field theory.

TETRAHEDRAL COMPLEXES

Here the situation is in many ways the reverse of the octahedral case. The ligands are now non-axial and their distribution around the central metal atom is shown in Fig. 17. Comparison with the shapes of the orbitals (Fig. 13) shows that, because of their closer proximity to the ligands, it is the non-axial orbitals (d_{xy}, d_{xz} and d_{yz}) which are destabilized most (Fig. 18). The d orbitals are now split into an upper d_ε or t_2 set of three and a lower d_γ or e set of two (in the t_2 and e symbolism the subscript g, used in the octahedral case, is omitted when referring to tetrahedral).

Unlike the d_γ set in the octahedral case, the d_ε orbitals, although now closest to the ligands, do not point directly at the ligands. For this reason, and also because there are fewer ligands, the tetrahedral orbital splitting, Δ_t, is less than the octahedral. In fact, for the same metal and ligands and the same internuclear distances

$$\Delta_t = \tfrac{4}{9}\Delta_0$$

Because of this relatively small orbital splitting, spin-pairing in tetrahedral compounds is rather unlikely. Although the diamagnetism of the d^4 complex,

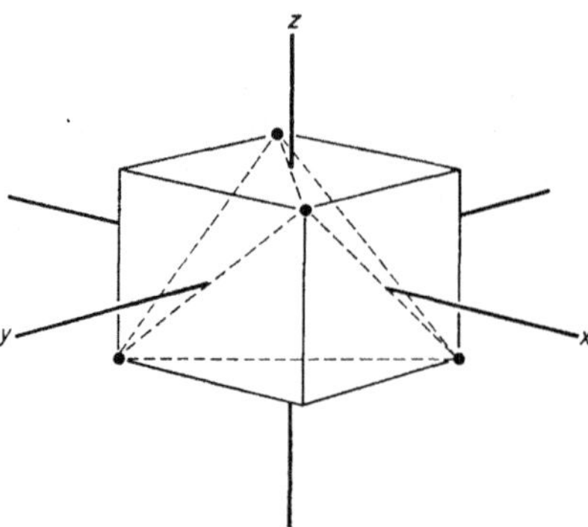

FIG. 17. In a tetrahedral complex the four ligands are non-axial.

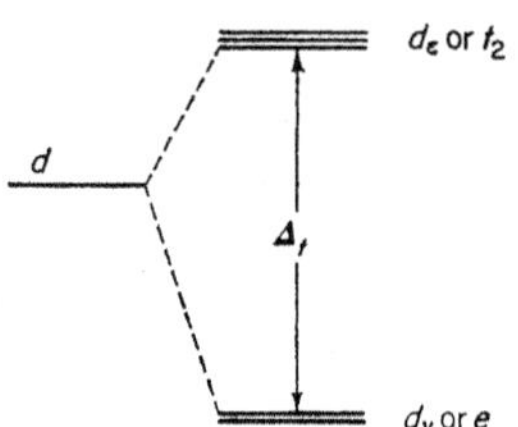

FIG. 18. The splitting of d orbitals in a tetrahedral complex. This is the reverse of that found in octahedral complexes.

TABLE IX. *Orbital contributions in octahedral and tetrahedral complexes of* d^x *ions*

No. of d electrons	Spin-paired oct.		Spin-free oct.		Spin-free tet.	
	Orb. cont.	*No* orb. cont.	Orb. cont.	*No* orb. cont.	Orb. cont.	*No* orb. cont.
1			d_ε^1			d_γ^1
2			d_ε^2			d_γ^2
3				d_ε^3	$d_\gamma^2 d_\varepsilon^1$	
4	d_ε^4			$d_\varepsilon^3 d_\gamma^1$	$d_\gamma^2 d_\varepsilon^2$	
5	d_ε^5			$d_\varepsilon^3 d_\gamma^2$		$d_\gamma^2 d_\varepsilon^3$
6		d_ε^6	$d_\varepsilon^4 d_\gamma^2$			$d_\gamma^3 d_\varepsilon^3$
7		$d_\varepsilon^6 d_\gamma^1$	$d_\varepsilon^5 d_\gamma^2$			$d_\gamma^4 d_\varepsilon^3$
8				$d_\varepsilon^6 d_\gamma^2$	$d_\gamma^4 d_\varepsilon^4$	
9				$d_\varepsilon^6 d_\gamma^3$	$d_\gamma^4 d_\varepsilon^5$	

$RbReCl_4$, was ascribed to spin-pairing it has recently been shown to arise instead from metal–metal bonding in a trimeric anion. If then the possibility of spin-pairing is ignored, the configurations d^3, d^4, d^8 and d^9 should retain some orbital contribution while in the remaining d configurations the orbital contribution should be completely quenched. This is an exact reversal of the spin-free, octahedral case (spin-free d^5 has no orbital angular momentum and can therefore never give rise to any orbital contribution).

Table IX summarizes the occurrence of orbital contribution in octahedral and tetrahedral transition metal complexes.

Other Stereochemistries

The orbital splitting in square planar complexes can be derived from the octahedral case by assuming a tetragonal distortion caused by withdrawing the two ligands on the z axis. In an octahedral complex the $d_z{}^2$ is the orbital

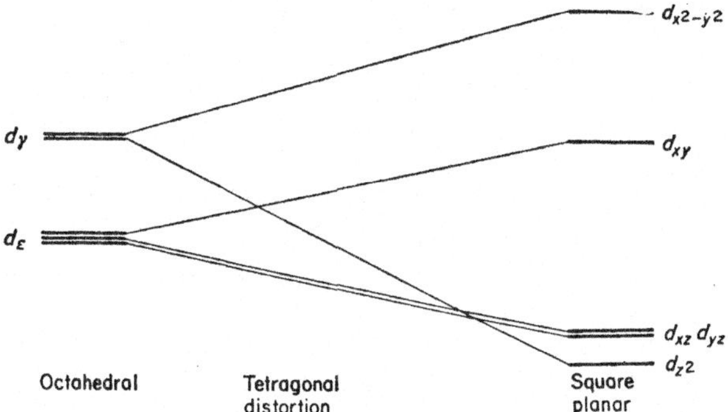

Fig. 19. The effect of an increasing tetragonal distortion in an octahedral complex, leading eventually to a square planar stereochemistry.

most strongly repelled by the ligands on the z axis, followed by the d_{xz} and d_{yz}. Removal of the ligands on the z axis will therefore stabilize these orbitals, the most marked effect being on the $d_z{}^2$ as shown in Fig. 19.

Whether or not the $d_z{}^2$ orbital actually lies lowest in a square planar complex will depend on the particular metal and ligands involved.

The orbital splittings for some other stereochemistries are summarized in Fig. 20.

Reported work on these stereochemistries is very sparse compared with the more highly symmetrical octahedral and tetrahedral arrangements and so exact predictions would be rather rash. However, it is evident that the lower the symmetry, the more the d orbitals lose their degeneracy, and the greater the likelihood of the orbital contribution being quenched.

It is worth noting that the compounds of the first row transition metals which have been most extensively examined are probably those of octahedral

Cr(III) (d^3), Fe(III) (d^5), Mn(II) (d^5), Ni(II) (d^8) and Cu(II) (d^9), none of which as it happens (except spin-paired d^5) are expected to give orbital contribution. The wide acceptance of the applicability of spin-only moments is thus to a very large extent coincidental.

The simple crystal field theory so far dealt with provides at least a qualitative reason for the applicability of spin-only moments in some cases and the occurrence of orbital contribution in others. It also indicates how, in principle, the presence or absence of orbital contribution may be taken to distinguish

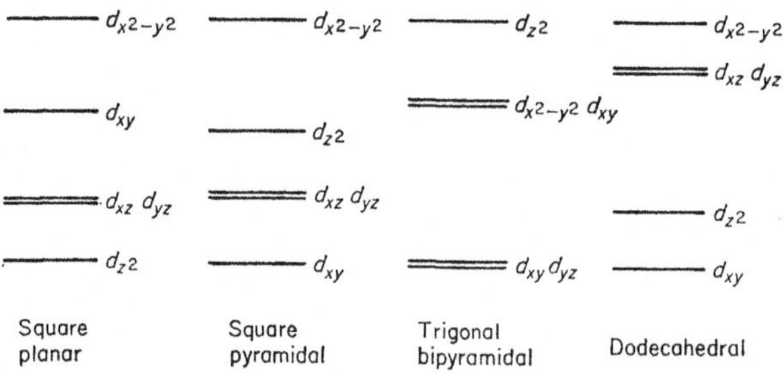

FIG. 20. The splitting of d orbitals by some non-cubic fields.

octahedral and tetrahedral shapes, since in no case does it appear that it should occur in both. However, there still remain instances (tetrahedral CoII is probably the most notable) which unfortunately do not fit into this simple scheme. Nor has it been shown how the temperature dependence of some moments might arise.

These deficiences arise because we have so far ignored completely the effect of spin-orbit coupling in transition metal compounds. In order to take account of this it is necessary to consider the situation from the point of view of the appropriate spectroscopic terms. This provides a more quantitatively useful, though less readily envisaged, approach.*

* For references for this chapter, see groups 3, 4, 6 and 13 on pp. 111 and 112.

IV. CRYSTAL FIELD THEORY

In the previous chapter spin-orbit coupling was completely ignored. This was because this additional effect is not easily included if the action of a ligand field is considered simply on the basis of electron occupancy of atomic orbitals. In order to take account of spin-orbit coupling it is necessary to look instead at the effect of the various interelectronic interactions on the spectroscopic terms of the ion. This view of crystal field theory is, in fact, the same approach as that already encountered in Chapter II but with the presence of the ligand field added.

The procedure to be adopted, therefore, is as follows.

1. The scheme of interelectronic interactions, arranged in order of their decreasing magnitudes, is drawn up.

2. The splitting of the spectroscopic terms into separate energy levels is worked out by the successive application of these interactions, ending with the effect of the applied magnetic field.

3. The bulk susceptibility is calculated by considering the thermal distribution of ions within the various levels.

The third step is facilitated by the use of the general equation for susceptibility due to Van Vleck. This equation will now be derived before dealing with the possible interaction schemes.

VAN VLECK'S FORMULA FOR SUSCEPTIBILITY

The change in energy of a magnetic dipole, such as a paramagnetic ion, due to the application of a magnetic field is

$$-\bar{\mu}H$$

where $\bar{\mu}$ is the projection of the moment of the ion *in the direction* of the field. The negative sign indicates a lowering in energy when the moment is aligned with the field. Differentiation gives

$$\frac{\partial E}{\partial H} = -\bar{\mu} \tag{1}$$

Now, in a completely general way, the energy of a particular level, i, of an ion in a magnetic field may be expressed as

$$E_i = E_{i(0)} + E_{i(1)} H + E_{i(2)} H^2 + \tag{2}$$

where $E_{i(0)}$ is the energy of level i in the absence of H, $E_{i(1)}$ is the coefficient of the first order Zeeman effect, and $E_{i(2)}$ is the coefficient of the second order Zeeman effect. (In cases where there is no magnetic moment associated with the level i then, of course, the first and second order coefficients will be zero.) Therefore, from (1) and (2)

$$\bar{\mu}_i = -E_{i(1)} - 2E_{i(2)} H - \tag{3}$$

The generality is increased by assuming that there are several different levels, each with a corresponding value of $\bar{\mu}$, and that they are thermally populated according to Boltzmann's distribution law. The values of χ_M and μ_e are then obtained by essentially the same procedure as was used in Chapter II.

$$I = \frac{N\bar{\mu}}{V} \quad \text{and} \quad \chi = \frac{I}{H \times \rho} = \frac{I}{H} \cdot \frac{V}{M}$$

$$\therefore \quad \chi_M = \frac{IV}{H} = \frac{N}{H} \cdot \bar{\bar{\mu}} \tag{4}$$

where N is Avogadro's number, V is the molar volume, M is the molecular weight, and $\bar{\bar{\mu}}$ is the average $\bar{\mu}_i$ for each molecule. However,

$$\bar{\bar{\mu}} = \frac{\sum \bar{\mu}_i n_i}{N}$$

and the Boltzmann distribution law states that

$$n_i = n_0 \exp\left(-\frac{E_i}{kT}\right)$$

where n_i is the number of molecules per mole in the level i

$$\therefore \quad \bar{\bar{\mu}} = \frac{n_0 \sum \bar{\mu}_i \exp\left(-\dfrac{E_i}{kT}\right)}{n_0 \sum \exp\left(-\dfrac{E_i}{kT}\right)} \tag{5}$$

From equation (2),

$$\exp\left(-\frac{E_i}{kT}\right) = \exp\left(\frac{-E_{i(0)} - E_{i(1)} H - E_{i(2)} H^2 \dots}{kT}\right)$$

$$= \exp\left(-\frac{E_{i(0)}}{kT}\right) \exp\left(\frac{-E_{i(1)} H - E_{i(2)} H^2 \dots}{kT}\right)$$

If it is assumed that $kT \gg E_{i(1)}H + E_{i(2)}H^2 + \ldots$ then

$$\exp\left(-\frac{E_i}{kT}\right) \simeq \left(1 - \frac{E_{i(1)}H}{kT}\right)\exp\left(-\frac{E_{i(0)}}{kT}\right) \qquad (6)$$

Substituting equations (3) and (6) in equation (5) gives:

$$\bar{\bar{\mu}} \simeq \frac{\sum(-E_{i(1)} - 2E_{i(2)}H)\left(1 - \dfrac{E_{i(1)}H}{kT}\right)\exp\left(-\dfrac{E_{i(0)}}{kT}\right)}{\sum\left(1 - \dfrac{E_{i(1)}H}{kT}\right)\exp\left(-\dfrac{E_{i(0)}}{kT}\right)} \qquad (7)$$

Equation (7) can be simplified because it is known, in practice, that for paramagnetic substances χ is independent of field strength.* Thus, because of equation (4), only terms up to H must be retained in the numerator and only terms independent of H in the denominator.

Also, since there is no permanent alignment of the magnetic moments in the absence of a magnetic field, it follows that the possible orientations of the moments in the direction of the applied field are exactly mirrored by similar orientation against the direction of the field. In other words, the result of the first order Zeeman effect is to split the state symmetrically into lower and upper levels. Therefore

$$\sum E_{i(1)} = 0$$

and so

$$\sum E_{i(1)}\exp\left(-\frac{E_{i(0)}}{kT}\right) = 0$$

With these simplifications equation (7) reduces to

$$\bar{\bar{\mu}} = \frac{H\sum\left(\dfrac{E_{i\,(1)}^2}{kT} - 2E_{i(2)}\right)\exp\left(-\dfrac{E_{i(0)}}{kT}\right)}{\sum \exp\left(-E_{i(0)}/kT\right)}$$

and from equation (4)

$$\chi_M = \frac{N\sum\left(\dfrac{E_{i\,(1)}^2}{kT} - 2E_{i(2)}\right)\exp\left(-\dfrac{E_{i(0)}}{kT}\right)}{\sum \exp\left(-\dfrac{E_{i(0)}}{kT}\right)} \qquad (8)$$

Note that in the summations degenerate levels are counted separately. The generality of equation (8) can perhaps be illustrated by applying it to a

---

* This is true unless the atomic magnetic moments approach perfect alignment with the applied field. For field strengths normally encountered this only occurs in the case of ferromagnetic materials and can be ignored for normal paramagnetic substances. This assumption is equivalent to saying that "saturation" effects are being ignored. (See Chapter V and the Appendix.)

free ion when the multiplet widths are large compared to kT. This was previously discussed in Chapter II. In this case there are $2J+1$ levels belonging to the ground state and, since the multiplet widths are large, no other states need be considered, i.e.

$$E_{i\,(0)} = 0 \quad \text{and} \quad \exp\left(-\frac{E_{i\,(0)}}{kT}\right) = 0$$

If, as before, terms in H^2 are neglected (i.e. assuming only a first order Zeeman effect), it follows that the energies of the $2J+1$ levels range from $-g\beta HJ$ to $+g\beta HJ$. The numerator of equation (8) becomes

$$\frac{g^2\beta^2}{kT}\left[J^2+(J-1)^2+\ldots 0 \ldots (-J)^2\right]$$

$$= \frac{2g^2\beta^2}{kT}\frac{J(J+1)(2J+1)}{6}$$

and the denominator $\qquad = (2J+1)$

$$\therefore \quad \chi_M = \frac{Ng^2\beta^2}{3kT}\,J(J+1)$$

which is as previously found and from which μ_e may be obtained by application of Langevin's formula.

STRENGTH OF LIGAND FIELDS

If Russell-Saunders coupling is assumed, the interelectronic interactions are taken in the order

$$s_i s_k > l_i l_k > s_i l_i$$

The effect of a ligand field acting on a metal ion in a co-ordination compound is considered in this scheme at the point appropriate to its magnitude in relation to the other interactions.

Four cases arise:

1. The ligand field being smaller than any of the other interactions. In this case the ion in the ligand field behaves virtually as a free ion. This is the case of large multiplet widths as exemplified by the lanthanides.

2. The ligand field being larger than the spin-orbit interaction but less than the others. This is the weak field case.

3. The ligand field being larger than both orbital and spin-orbit coupling but less than spin coupling. This is the medium field case and corresponds to the breakdown of Hund's second rule. The ground state is that with

maximum multiplicity but the electrons are confined to "private" d_ε and d_γ orbitals.

4. The ligand field being larger than all other interactions. This is the strong field case and corresponds to the breakdown of Hund's first rule. The ground state is now not necessarily that with maximum multiplicity and, for d^4, d^5, d^6, and d^7 ions, spin pairing results.

In dealing with transition metal complexes case 1 can be ignored. Indeed it is often sufficient to deal only with cases 2 and 4.

Weak field complexes are the high-spin, spin-free or outer orbital complexes of other classifications, and strong field complexes are the low-spin, spin-paired or inner orbital types. Although most complexes can conveniently be considered in one or other of these categories, it is important to realize that no abrupt change between them is to be expected.

WEAK FIELD CASE

$$s_i s_k > l_i l_k > \text{L.F.} > s_i l_i$$

The ground term of the ion is deduced by considering the spin and orbital coupling in exactly the same way as for the free ion. The ground terms for the d^x ions have already been given (p. 20) and are S, D or F terms corresponding to $L = 0$, 2 or 3. Just as the multiplicity, or spin degeneracy, of a term is given by $2S+1$, so its orbital degeneracy is given by $2L+1$ which is 1, 5 and 7 for S, D and F terms. The effect of the ligand field is to lift at least some of this orbital degeneracy of the ground term but, being electrostatic in nature, it can have no direct effect on the spin degeneracy.

To avoid confusion it must be made plain at the outset that the expression "orbital degeneracy" will be used only to refer to the number of possible orientations of L and *not* to the number of degenerate orbitals. The same applies to such expressions as "orbital singlet", "orbital doublet", etc., which will be used in the ensuing discussion, but not to "orbital splitting" which has already been used in reference to the way in which the energies of the orbitals are affected by a ligand field.

In order to understand the effect of a ligand field on the spectroscopic terms of an ion, it is helpful to draw analogies with the corresponding effect on the atomic orbitals of the ion. Orbitals, the degeneracy ($= 2l+1$) of which arises from different m_l values, are designated by lower case letters: spectroscopic terms, the degeneracy ($= 2L+1$) of which arises from different M_L values, are designated by capital letters. In both cases an octahedral field is distinguished from a tetrahedral by the addition of a subscript g to the orbital and term symbols of the former.

S Terms

Just as an s orbital, possessing spherical symmetry and being non-degenerate ($l = 0$), cannot be split by a ligand field, so an S term likewise remains unsplit. It is, however, given the symbol A_1, in conformity with the symbol used for the orbital singlet obtained in the splitting of an F term.

D Terms

The fivefold degeneracy of d orbitals, as has been shown, is lifted by a cubic ligand field to give doublet e and triplet t_2 sets of orbitals. Similarly a D term is split into E and T_2 terms, separated by Δ, or 10 Dq. There is an important difference between orbital splitting and term splitting which is illustrated in Fig. 21. In the former only a change of stereochemistry (octahedral to tetra-

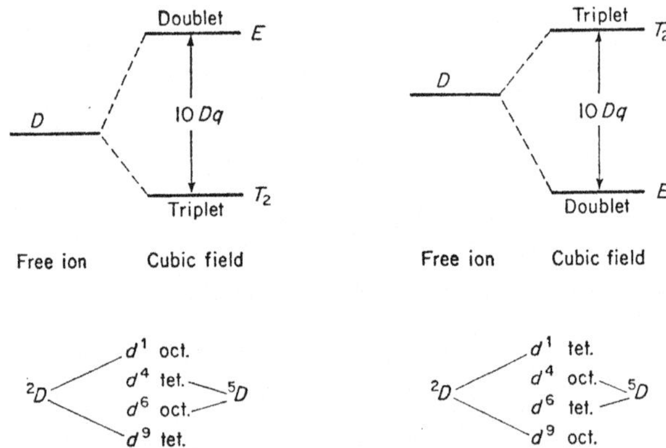

FIG. 21. The splitting of free ion D terms by cubic fields.

hedral) causes an inversion of the splitting, but in the latter this can also be produced by a change of electronic configuration from d^x to d^{10-x}.

F Terms

Although the splitting of the seven f orbitals by a cubic ligand field has not been dealt with, it can be shown to produce two triplet sets of orbitals and a singlet (reference 10). In the same way the sevenfold degeneracy of an F term is lifted to give two triplet terms and a singlet, given the symbols T_1, T_2 and A_2, with separations of 8 Dq and 10 Dq, as shown in Fig. 22. As with D terms the splitting is inverted by a change of electronic configuration from d^x to d^{10-x}.

The stereochemical and configurational inversion of term splitting is very well illustrated by the Orgel diagrams (Figs. 23 and 24) for D and F terms.

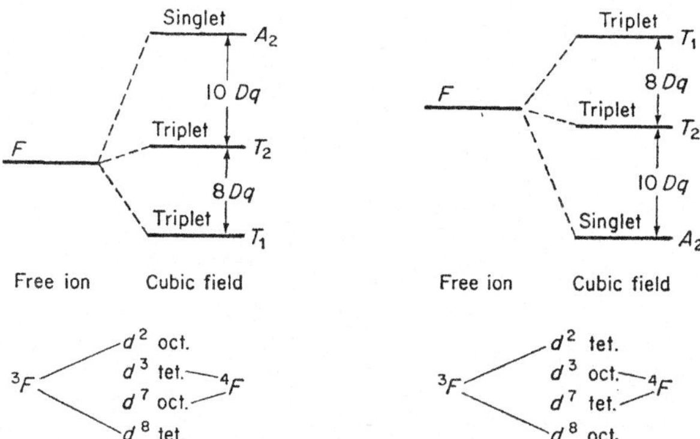

FIG. 22. The splitting of free ion F terms by cubic fields.

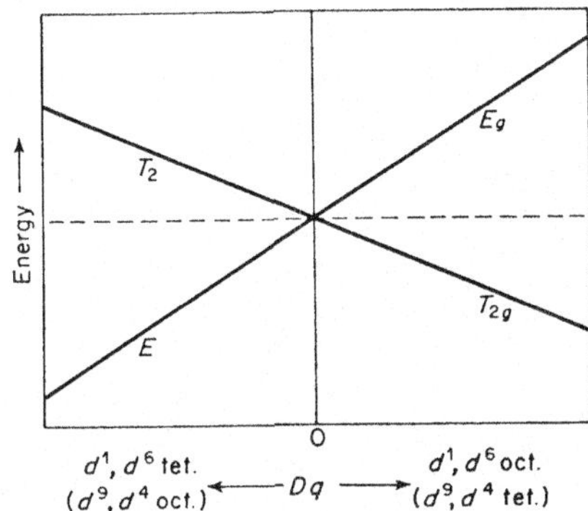

FIG. 23. Orgel diagram showing the splitting of a D term in an octahedral field (to the right) and a tetrahedral field (to the left). For the d^{10-x} configurations the octahedral field is to the left and the tetrahedral to the right (the g subscripts being reversed also).

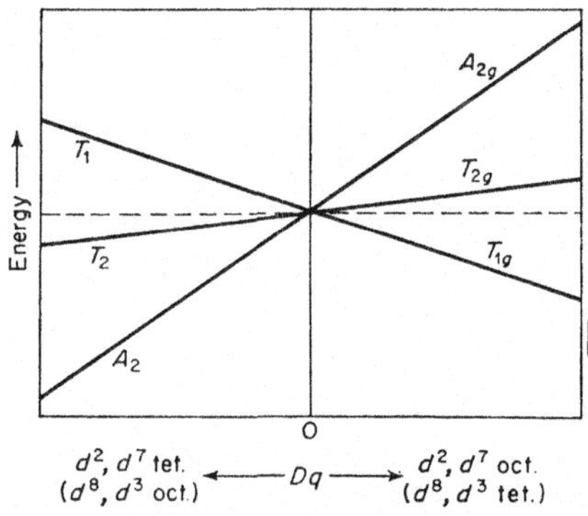

FIG. 24. Orgel diagram showing the splitting of an F term in octahedral and tetrahedral fields. Although the variation of energy with Dq is represented as linear, this assumes the terms to be independent of all others. In practice, interactions with higher terms are likely to occur causing deviations from this simple representation.

STRONG FIELD CASE

$$\text{L.F.} > s_i s_k > l_i l_k > s_i l_i$$

The ligand field is now so large that it may result in a ground term of spin multiplicity less than that of the ground term of the free ion.

Kotani has treated this problem in the manner just adopted for the weak field case; namely by taking the various interactions in the order of their decreasing magnitudes. It is possibly seen more clearly, however, if, instead, the effect of a gradually increasing ligand field on the terms of the free ion is noted. This has the advantage that both weak and strong field extremes are included, showing the connexion between the two. Only octahedral fields

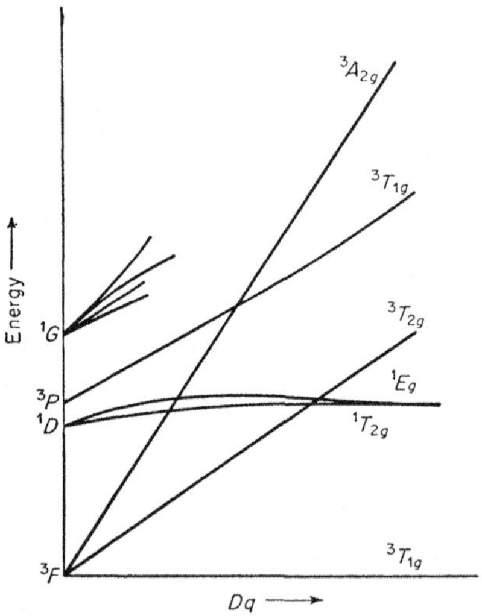

Fig. 25. Energy diagram for d^2 ions.

will be considered since, as previously mentioned, no case is known in which a tetrahedral field may be classified as "strong". The results are shown in Figs. 25–31. These are simplified forms of the diagrams originally drawn up by Tanabe and Sugano. d^1 and d^9 configurations are not included since they give rise only to a D term which is exactly the same (except in the magnitude of its splitting) in weak or strong octahedral fields.

d^2, d^3 and d^8 configurations retain the same ground term in strong fields as in weak but for the other configurations, d^4 to d^7, a strong ligand field has the effect of forcing a term of lower multiplicity below the free ion ground term.

The first conclusion to be drawn is that, whether the ligand field is weak or strong, the ground term of any transition metal ion is always an orbital singlet

(A, from S or F) or doublet (E, from D) or triplet (T, from D or F). The orbital doublet E, arising from a D term is, like the d_γ pair of orbitals, known as a "non-magnetic" doublet because it is associated with no orbital angular

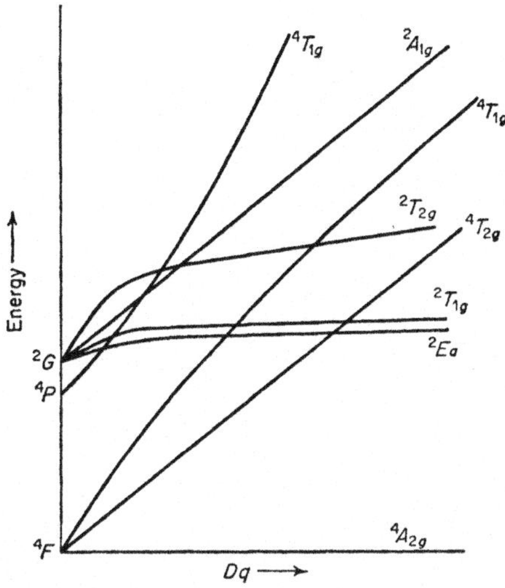

FIG. 26. Energy diagram for d^3 ions.

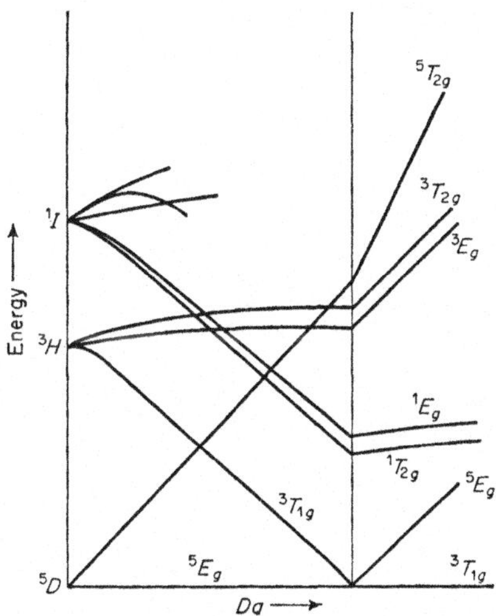

FIG. 27. Energy diagram for d^4 ions.

momentum and hence gives rise to no orbital contribution. Thus, for ions in which a singlet A or a doublet E term lies lowest, no orbital contribution is expected but, if a triplet T term lies lowest, orbital contribution should occur leading to a moment in excess of $\mu_{\text{s.o.}}$. The conclusions to be drawn from this about stereochemistry are just those which were drawn in the preceding chapter. To refine them it is necessary to look at the effect of spin-orbit

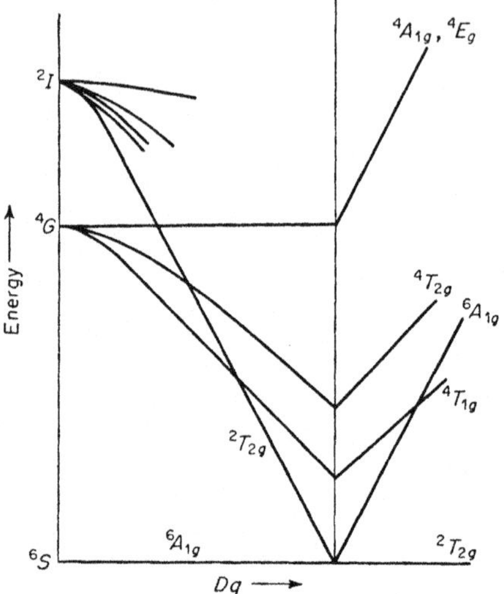

FIG. 28. Energy diagram for d^5 ions.

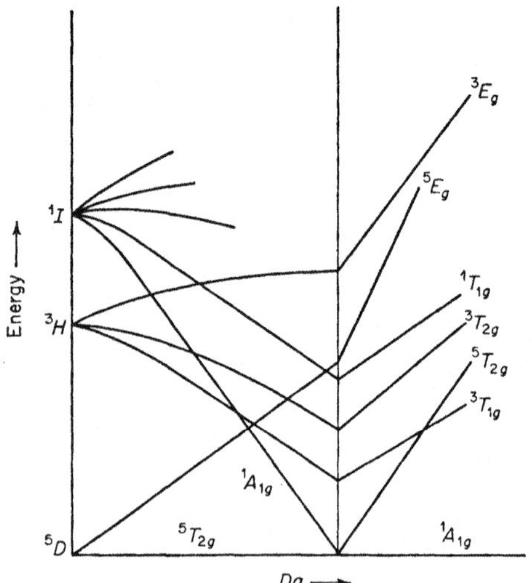

FIG. 29. Energy diagram for d^6 ions.

coupling on the A, E and T terms, but the situation might possibly be clearer if first a correlation is attempted between term splitting and the electron configurations of transition metal ions.

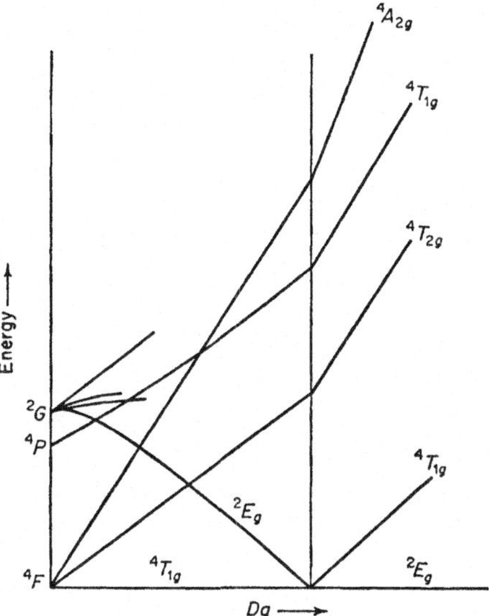

FIG. 30. Energy diagram for d^7 ions.

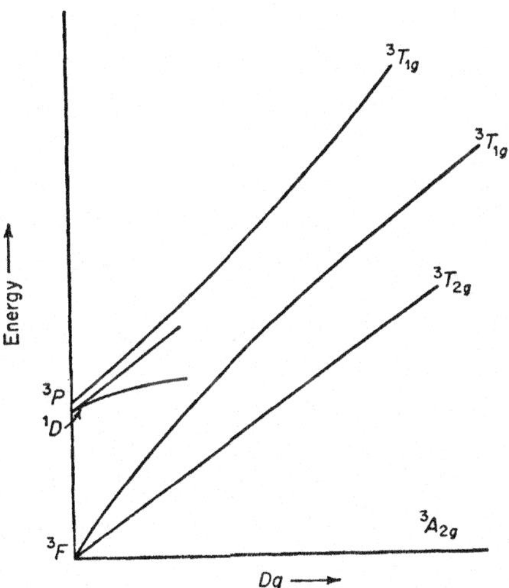

FIG. 31. Energy diagram for d^8 ions.

Figs. 25 to 31 must not be assumed to be drawn accurately to scale (the scales vary slightly for different ions of the same configuration) and, for simplicity, many of the higher terms have been omitted. When a higher term falls below the initial ground term it is conventional to maintain this new ground term coincident with the abscissa, necessitating the bending upward of all other terms by a corresponding amount.

C

TERM SPLITTING AND ELECTRON CONFIGURATION

The effect of a ligand field is to remove some of the free ion degeneracy of atomic orbitals, causing the lower ones to be preferentially occupied by electrons. Occupation of the higher orbitals may occur, but only if the requisite amount of energy is available. The spectroscopic terms of an ion describe these different configurations and their energies are the energies of the configurations, the lowest term corresponding to the most stable configuration. It would be untrue to say that each term corresponds to a unique configuration since a given term may contain an admixture of another configuration and a particular configuration may give rise to more than one term. This is because, for a given electron configuration, the orbital angular momentum vectors may couple in more than one way, each corresponding to a different spectroscopic term. Nevertheless, it is helpful in comparing the two approaches to correlate each term with the configuration which it most nearly describes (Table X).

d^1. The free ion ground term, ^2D, with an orbital degeneracy of 5, is split by an octahedral field into a lower orbital triplet and an upper orbital doublet. The triplet corresponds to d_ε^1 when the electron can occupy any of the three degenerate d_ε orbitals. The doublet similarly may be taken to correspond to d_γ^1.

d^9. This is easily obtained by imagining a single "positive hole" which is the exact reverse of the d^1 configuration. Thus an octahedral field produces a lower doublet and an upper triplet.

It follows that octahedral d^1 and tetrahedral d^9 are similar, as are tetrahedral d^1 and octahedral d^9. By using the "positive hole" formalism this general correspondence between octahedral d^x and tetrahedral d^{10-x} is seen to occur throughout.

d^2. The free ion ground term ^3F, with an orbital degeneracy of 7, is split by an octahedral field into two triplets and a singlet, one of the triplets being lowest. This corresponds to d_ε^2 when the two electrons can be arranged in the degenerate d_ε orbitals in three ways (remembering Hund's rule that the spins must not be paired). The singlet is highest and may be assumed to correspond to the promotion of both electrons to the d_γ orbitals where there is only one possible arrangement.

d^8. Again on the basis of two positive holes, this is the reverse of the d^2 configuration.

d^3. This is another F ground term and an octahedral field gives a lower singlet term since the three electrons in the d_ε orbitals can only be arranged in one way. The upper triplet corresponds to the promotion of two unpaired electrons to the d_γ orbitals, when the remaining electron may be in any one of the three d_ε orbitals.

d^7. Three positive holes give the reverse of the d^3 configuration.

d^4. This is another D ground term and an octahedral field gives a lower doublet corresponding to the single occupation of either of the two d_γ orbitals, the lower d_ε orbitals being singly occupied in both cases. The upper triplet corresponds to the promotion of two unpaired electrons to the d_γ set when the remaining two electrons may be arranged in three ways in the d_ε orbitals.

d^6. Four positive holes give the reverse of the d^4 configuration.

d^5. The free ion ground term ^6S has no orbital degeneracy, there being only

TABLE X. *Approximate correlation of term splittings and electron configurations for* d^x *ions in weak octahedral fields*

d^1	^2D	d^2	^3F	d^3	^4F	d^4	^5D
		d_γ^2	$^3A_2$	$d_\varepsilon^1 d_\gamma^2$	$^4T_1$		
d_γ^1	^2E					$d_\varepsilon^2 d_\gamma^2$	$^5T_2$
		$d_\varepsilon^1 d_\gamma^1$	$^3T_2$	$d_\varepsilon^2 d_\gamma^1$	$^4T_2$		
d_ε^1	$^2T_2$					$d_\varepsilon^3 d_\gamma^1$	^5E
		d_ε^2	$^3T_1$	d_ε^3	$^4A_2$		

d^6	^5D	d^7	^4F	d^8	^3F	d^9	^2D
		$d_\varepsilon^3 d_\gamma^4$	$^4A_2$	$d_\varepsilon^4 d_\gamma^4$	$^3T_1$		
$d_\varepsilon^3 d_\gamma^3$	^5E					$d_\varepsilon^5 d_\gamma^4$	$^2T_2$
		$d_\varepsilon^4 d_\gamma^3$	$^4T_2$	$d_\varepsilon^5 d_\gamma^3$	$^3T_2$		
$d_\varepsilon^4 d_\gamma^2$	$^5T_2$					$d_\varepsilon^6 d_\gamma^3$	^2E
		$d_\varepsilon^5 d_\gamma^2$	$^4T_1$	$d_\varepsilon^6 d_\gamma^2$	$^3A_2$		

For tetrahedral fields all the above terms are inverted.

one way of arranging five unpaired electrons in five orbitals, and the term is unaffected by a ligand field.

The fact that d^5 gives a singlet ground term suggests an alternative to the "positive hole" formalism adopted for configurations of $x > 5$; this is to neglect the first five electrons, which contribute nothing to the orbital degeneracy, i.e. $d^x \equiv d^{x-5}$. The pairs of configurations which emerge are: d^1 and d^6, d^2 and d^7, d^3 and d^8, d^4 and d^9, the members of which are similar to each other in octahedral and also in tetrahedral fields.

In strong octahedral fields the configurations d^4, d^5, d^6 and d^7 have ground terms $^3T_{1g}$ $^2T_{2g}$ $^1A_{1g}$ and ^2E_g (Figs. 27–30) which correspond to d_ε^4 d_ε^5 d_ε^6 and $d_\varepsilon^6 d_\gamma^1$. These correlations are readily explained along the same lines as above and correspond exactly to the spin-pairing derived on the basis of orbital splitting.

The most important point emerging from these correlations is that those cases in which an orbital singlet or doublet lies lowest are precisely those which, from the last chapter, are not expected to give any orbital contribution. Conversely, when an orbital triplet lies lowest the corresponding electronic configurations are those which do give rise to orbital contribution. It is again evident that, for complexes of lower symmetry, further splitting is likely and this will probably reduce still more the number of cases in which an orbital contribution can be expected.

It is now opportune to proceed with the examination of the effect of spin-orbit coupling on A, E and T terms, and to see the way in which the previous qualitative predictions about orbital contribution are thereby modified.

SPIN-ORBIT COUPLING ON A AND E TERMS

To a first approximation these terms give no orbital contribution. However, if there is a T term of the same multiplicity at a higher energy, it is possible for spin-orbit coupling to "mix" some of this into the ground term, so introducing a certain amount of orbital angular momentum into the latter. The reason for this is that, though a ligand field is able to quench orbital angular momentum, it has no effect on spin angular momentum. Consequently, if the two are coupled by spin-orbit coupling, the ligand field is unable to effect a perfect separation of terms of the same multiplicity on the basis of their differing orbital angular momentum. The ground term may, therefore, retain some orbital angular momentum, which is equivalent to saying that it is not a pure A or E term.

There are two cases in which the ground term is A_1. These are the weak field octahedral d^5 and the strong field octahedral d^6 configurations. In the former there is no higher T term of the same multiplicity and it is therefore expected to give a spin-only moment of 5·92 B.M., independent of temperature. The latter is diamagnetic anyway.

A_2 and E terms arising from F and D terms respectively (also the ^2Eg term of strong field octahedral d^7 arising from the ^2G term) must, of necessity, be accompanied by a higher T term of the same multiplicity. For these the "mixing in" effect is expressed by the formula:

$$\mu_e = \mu_{\text{s.o.}} \left(1 - \alpha\, \frac{\lambda}{10\, Dq}\right) \tag{9}$$

where α is 2 for an E, and 4 for an A_2 term, $10\, Dq$ is the energy separating the interacting terms, and λ is the spin-orbit coupling constant for the terms involved.

In the free ion, spin-orbit coupling lifts the degeneracy of the terms, splitting them into "states". The extent of this splitting is proportional to the scalar products of the spin and orbital angular momentum vectors and λ is the constant of proportionality. It refers to the particular term and is related to the spin-orbit coupling constant, ζ, of an individual electron (ζ is a property of the configuration) by:

$$\lambda = \pm \frac{\zeta}{2S} \tag{10}$$

This is particularly important in dealing with the strong field case, when the value of S to be used is that appropriate to the term lying lowest. ζ is always

TABLE XI. *Free ion values* (cm^{-1}) *of spin-orbit coupling constants for first row transition metal ions*

Ligand field:			Weak oct.		Strong oct.		Weak tet.	
Ion	ζ	No. of d electrons	Ground term	λ	Ground term	λ	Ground term	λ
Ti^{3+}	155	1	$^2T_{2g}$	$+155$	$^2T_{2g}$	$+155$	^2E	$+155$
V^{3+}	210	2	$^3T_{1g}$	$+105$	$^3T_{1g}$	$+105$	$^3A_2$	$+105$
V^{2+}	170	3	$^4A_{2g}$	$+57$	$^4A_{2g}$	$+57$	$^4T_1$	$+57$
Cr^{3+}	275	3	$^4A_{2g}$	$+92$	$^4A_{2g}$	$+92$	$^4T_1$	$+92$
Cr^{2+}	230	4	^5E_g	$+58$	$^3T_{1g}$	-115	$^5T_2$	$+58$
Mn^{3+}	355	4	^5E_g	$+89$	$^3T_{1g}$	-178	$^5T_2$	$+89$
Mn^{2+}	300	5	$^6A_{1g}$	—	$^2T_{2g}$	-300	$^6A_1$	—
Fe^{3+}	460	5	$^6A_{1g}$	—	$^2T_{2g}$	-460	$^6A_1$	—
Fe^{2+}	400	6	$^5T_{2g}$	-100	$^1A_{1g}$	—	^5E	-100
Co^{3+}	580	6	$^5T_{2g}$	-145	$^1A_{1g}$	—	^5E	-145
Co^{2+}	515	7	$^4T_{1g}$	-172	^2E_g	-515	$^4A_2$	-172
Ni^{3+}	715	7	$^4T_{1g}$	-238	^2E_g	-715	$^4A_2$	-238
Ni^{2+}	630	8	$^3A_{2g}$	-315	$^3A_{2g}$	-315	$^3T_1$	-315
Cu^{2+}	830	9	^2E_g	-830	^2E_g	-830	$^2T_2$	-830

positive and the $+$ sign in equation (10) applies to shells up to half full. The $-$ sign applies to shells which are more than half full, as can be understood if these configurations are treated on the "positive hole" basis, since this inverts the energies of the states produced when spin-orbit coupling splits the terms. The shells are the complete sets of d orbitals in weak fields and the d_ε sets in strong fields.

Table XI lists the values of ζ for di- and tri-valent ions of the first row transition elements. It can be seen that they increase from left to right along the row.

It is to be expected that ζ for an ion in a complex will differ from that for the free ion, but the difference is likely to be less than the experimental error introduced in the usual methods employed to measure μ_e.

An alternative way of visualizing the effect of spin-orbit coupling on μ_e is to regard it as a change in the value of the g factor

$$\mu_{\text{s.o.}} = g\beta\sqrt{S(S+1)}$$

where $g = 2$.

$$\therefore \quad \mu_e = g\beta\sqrt{S(S+1)}$$

where

$$g = 2\left(1 - \alpha\,\frac{\lambda}{10\,Dq}\right)$$

Thus A and E terms are regarded as pseudo S terms in which the orbital contribution is taken account of by the value assigned to g. The usefulness of this will be appreciated later in the discussion of intramolecular antiferro-magnetism.

Two important results emerge.

1. For A and E terms of weak field d^1 to d^4 configurations, λ is positive and should lead to a reduction in the moment below the spin-only value. For A and E terms of weak field d^6 to d^9 configurations, λ is negative and should lead to an increase in the moment above the spin-only value, the effect being more pronounced because λ is larger.

The only relevant strong field case is d^7, for which $2S = 1$. $\therefore \lambda = -\zeta$, leading to a moment above 1·73 B.M.

2. Since the difference in energy between the ground term and the term being mixed into it is very much larger than kT (perhaps 10 000 cm^{-1} as against 200 cm^{-1} near room temperature), the contribution to the moment is independent of temperature.

Taking $\text{Ni}(\text{H}_2\text{O})_6{}^{2+}$ as an example, 10 Dq obtained spectroscopically is 8 900 cm^{-1} and $\lambda = -315$ cm^{-1}. Since the ground term is $^3A_{2g}$, $\alpha = 4$ leading to

$$\mu_e = 2\cdot83\left(1 + \frac{1\,260}{8\,900}\right) = 3\cdot23 \text{ B.M.}$$

This agrees well with experimental values for the hexahydrate salts of Ni(II).

For first row ions in general the moments are increased by about 0·2–0·4 B.M. by this effect.

SPIN-ORBIT COUPLING ON T TERMS

Because these terms retain some orbital angular momentum they give rise to appreciable orbital contribution. However, spin-orbit coupling removes some of their degeneracy and a quantitative estimation of the orbital contribution is only possible if the actual magnitudes of the splittings are known. Their evaluation requires a more comprehensive treatment than can be attempted here but, if they are assumed, the resulting equations for χ_M and μ_e can be found by application of equation (8). As an indication of the general method, this will now be performed for the case of the d^1 configuration in an octahedral field (Fig. 32). The free ion ^2D term has a total degeneracy (spin $\times$ orbital $= (2S+1)(2L+1)$) of 10. The octahedral field removes some of this by splitting the term into an upper ^2E_g and a ground $^2T_{2g}$ term with total degeneracies of 4 and 6 respectively. The degeneracy of the ground term is then lifted successively by spin-orbit coupling, the first order Zeeman effect of

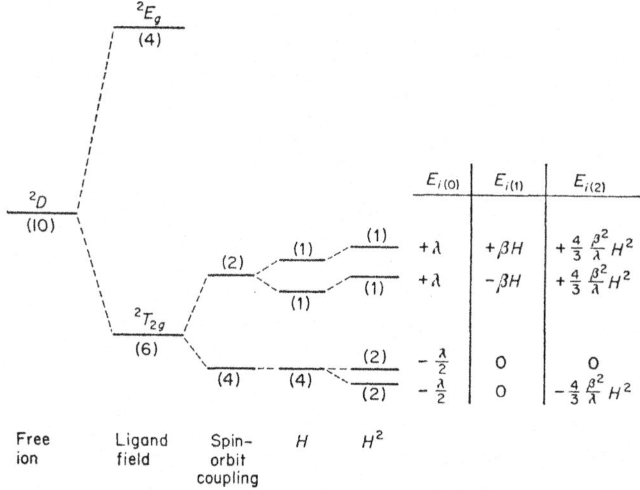

FIG. 32. The successive splitting of the ^2D ground term of a d^1 ion in an octahedral field. The degeneracy of each level is given in parentheses and the energies, E_i, are shown on the right.

an applied magnetic field (a magnetic field must be applied if the ion's magnetic properties are to be measured!), and finally by the second order Zeeman effect.

Insertion in equation (8) of the values of E given in Fig. 32 gives

$$\chi_M = \frac{N\left[2\left(\dfrac{\beta^2}{kT}-\dfrac{8\beta^2}{3\lambda}\right)\exp\left(-\dfrac{\lambda}{kT}\right)+0+2\left(0+\dfrac{8\beta^2}{3\lambda}\right)\exp\left(+\dfrac{\lambda}{2kT}\right)\right]}{2\exp\left(-\dfrac{\lambda}{kT}\right)+4\exp\left(+\dfrac{\lambda}{2kT}\right)}$$

Multiplying throughout by $\frac{1}{2}\exp(-x/2)$ where $x = \lambda/kT$

$$\chi_M = \frac{N\left[\left(\dfrac{\beta^2}{kT} - \dfrac{8\beta^2}{3\lambda}\right)\exp\left(-\dfrac{3x}{2}\right) + \dfrac{8\beta^2}{3\lambda}\right]}{\exp\left(-\dfrac{3x}{2}\right) + 2}$$

$$= \frac{N\beta^2}{3kT}\left[\frac{\left(3 - \dfrac{8}{x}\right)\exp\left(-\dfrac{3x}{2}\right) + \dfrac{8}{x}}{\exp\left(-\dfrac{3x}{2}\right) + 2}\right]$$

but

$$\chi_M = \frac{N\beta^2\mu_e^2}{3kT}$$

$$\therefore \quad \mu_e^2 = \frac{\left(3 - \dfrac{8}{x}\right)\exp\left(-\dfrac{3x}{2}\right) + \dfrac{8}{x}}{\exp\left(-\dfrac{3x}{2}\right) + 2}$$

$$\therefore \quad \mu_e^2 = \frac{8 + (3x - 8)\exp\left(-\dfrac{3x}{2}\right)}{x\left[2 + \exp\left(-\dfrac{3x}{2}\right)\right]} \tag{11}$$

Equation (11) applies equally in this case to weak or strong octahedral fields and also to a d^9 ion in a tetrahedral field, with the proviso in the last case that the sign of λ is negative.

The same general procedure can be adopted for the other configurations with a T ground term, leading to the results summarized in Table XII.

The graphic representations of the results in Table XII are given in Figs. 33–36. The interpretation of these equations and the curves drawn from them is that they describe the way in which the spin and orbital angular momenta of the ions reinforce or cancel each other depending on λ and T. This is rather complicated in most of the cases but in some the limiting values of μ_e when $\dfrac{kT}{|\lambda|} \to 0$ and $\dfrac{kT}{|\lambda|} \to \infty$ correspond to quite simple situations.*

For the octahedral d^1 configuration the magnetic moments associated with $L = 1$ and $S = \frac{1}{2}$ effectively oppose each other ($\mu_e = 0$) when $\dfrac{kT}{|\lambda|} = 0$, but

---

* These are easily worked out by the usual approximations, $\lim\limits_{x \to 0} \exp(x) = 1 + x$ and $\lim\limits_{x \to \infty} \exp(-x) = 0$, if the sign of λ is borne in mind and the appropriate rearrangements of the equations are performed.

TABLE XII. *Magnetic moments of transition metal ions with ground T terms in cubic fields*

C*

Configuration	λ	Ground term	$\mu_e{}^2$
			WEAK FIELD
d^1 oct.	$+\zeta$		
d^9 tet.	$-\zeta$	$^2T_2$	$\dfrac{8+(3x-8)\exp\left(-\dfrac{3x}{2}\right)}{x\left[2+\exp\left(-\dfrac{3x}{2}\right)\right]}$
d^2 oct.	$+\dfrac{\zeta}{2}$		
d^8 tet.	$-\dfrac{\zeta}{2}$	$^3T_1$	$\dfrac{3\left[0.625x+6.8+(0.125x+4.09)\exp(-3x)-10.89\exp\left(-\dfrac{9x}{2}\right)\right]}{x\left[5+3\exp(-3x)+\exp\left(-\dfrac{9x}{2}\right)\right]}$
d^6 oct.	$-\dfrac{\zeta}{4}$		
d^4 tet.	$+\dfrac{\zeta}{4}$	$^5T_2$	$\dfrac{3\,[28x+9.33+(22.5x+4.17)\exp(-3x)+(24.5x-13.5)\exp(-5x)]}{x\,[7+5\exp(-3x)+3\exp(-5x)]}$
d^7 oct.	$-\dfrac{\zeta}{3}$		
d^3 tet.	$+\dfrac{\zeta}{3}$	$^4T_1$	$\dfrac{3\left[3.15x+3.92+(2.84x+2.13)\exp\left(-\dfrac{15x}{4}\right)+(4.7x-6.05)\exp(-6x)\right]}{x\left[3+2\exp\left(-\dfrac{15x}{4}\right)+\exp(-6x)\right]}$
			STRONG FIELD
d^1 oct.	$+\zeta$		
d^5 oct.	$-\zeta$	$^2T_2$	Same as weak field d^1 oct.
d^2 oct.	$+\dfrac{\zeta}{2}$		
d^4 oct.	$-\dfrac{\zeta}{2}$	$^3T_1$	$\dfrac{3\,[5x+15+(x+9)\exp(-2x)-24\exp(-3x)]}{2x\,[5+3\exp(-2x)+\exp(-3x)]}$

$x = \lambda/kT$. Values of ζ and hence of λ may be obtained from Table XI.

they progressively align themselves reaching the spin only value of 1·73 B.M. when $\dfrac{kT}{|\lambda|} \simeq 1$. Finally, when $\dfrac{kT}{|\lambda|} \to \infty$, $\mu_e = \sqrt{5}$ B.M. corresponding to $\sqrt{4S(S+1)+L(L+1)}$.

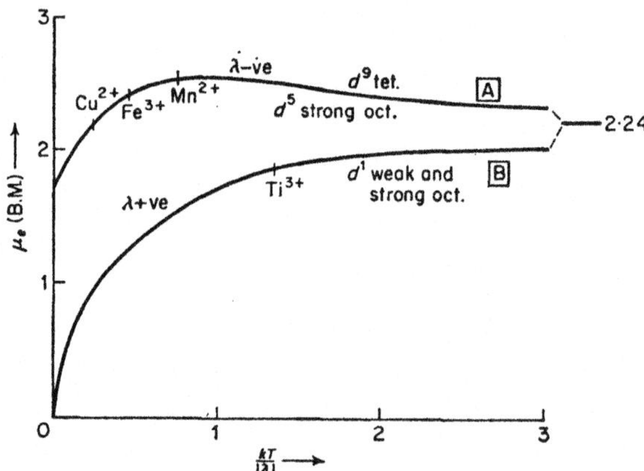

FIG. 33. The variation of magnetic moment with temperature for $^2T_2$ terms. A. $\lambda-$ve, i.e. d^9 in tetrahedral and d^5 in strong octahedral fields. B. $\lambda+$ve, i.e. d^1 in strong or weak octahedral fields. The moments of individual ions are marked for $T = 300°$K using the free ion values of λ.

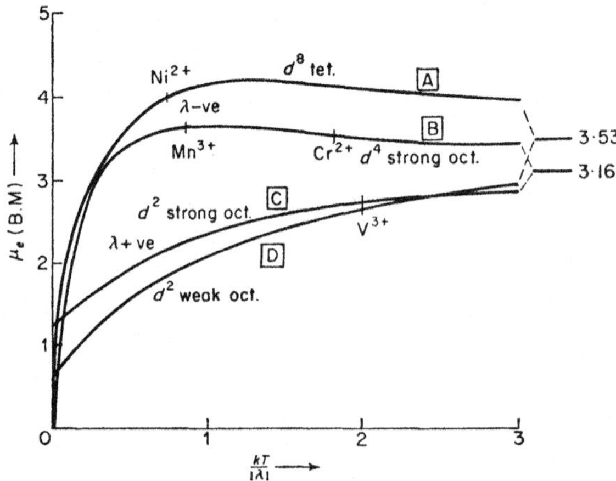

FIG. 34. The variation of magnetic moment with temperature for $^3T_1$ terms. A. $\lambda-$ve, i.e. d^8 in a tetrahedral field. B. $\lambda-$ve, i.e. d^4 in a strong octahedral field. C. $\lambda+$ve, i.e. in a strong octahedral field. D. $\lambda+$ve, i.e. d^2 in a weak octahedral field.

Similarly, for the octahedral strong field, d^4 configuration, the moments associated with $L = 2$ and $S = 1$ effectively cancel when $\dfrac{kT}{|\lambda|} = 0$. However, in this case the angular momenta never become completely aligned and, when $\dfrac{kT}{|\lambda|} \to \infty$, $\mu_e \, (= \sqrt{10}\ \text{B.M.})$ corresponds only to the alignment of the moments associated with $L = 1$ and $S = 1$.

An important point to be remembered in comparing these calculations with

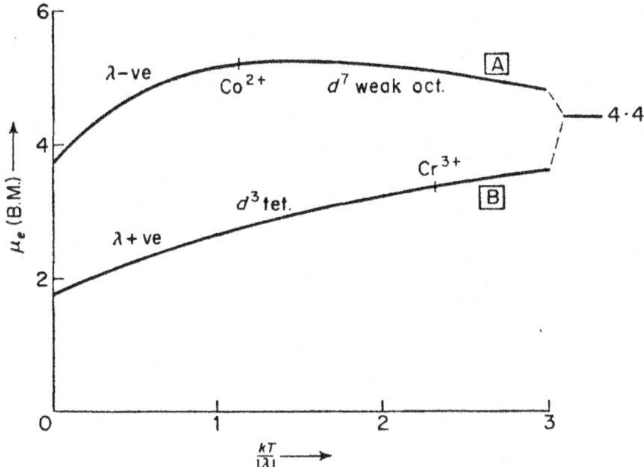

FIG. 35. The variation of magnetic moment with temperature for $^4T_1$ terms. A. $\lambda-$ve, i.e. d^7 in a weak octahedral field. B. $\lambda+$ve, i.e. d^3 in a tetrahedral field.

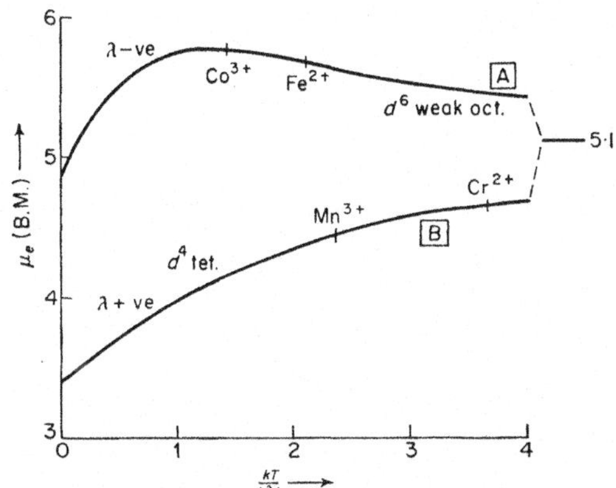

FIG. 36. The variation of magnetic moment with temperature for $^5T_2$ terms. A. $\lambda-$ve, i.e. d^6 in a weak octahedral field. B. $\lambda+$ve, i.e. d^4 in a tetrahedral field.

experimentally observed moments is that only the extremes of weak and strong fields have been treated; medium fields have been completely ignored. This is unlikely to be serious for tetrahedral complexes since for them the fields are usually weak. Nor does it matter when the T ground term arises from a D term (d^1 and d^6 oct.) because, in fact, then the weak and medium fields give the same result.

However, when the ground T term is produced from an F term (d^2 and d^7 oct., neglecting tet. cases) some disparity is likely since a medium field gives an appreciably different result. The most important instances of this are the octahedral ions V^{3+} and Co^{2+}. More detailed examination suggests that, for the former, moments rather higher and, for the latter, rather lower are likely.

The most important general conclusion to be drawn about the effect of spin-orbit coupling is that *those molecules in which there is no orbital contribution in the ground term* (A *and* E) *give rise to moments independent of temperature, while those molecules in which there is orbital contribution in the ground term* (T) *give moments which are dependent on temperature.*

The temperature dependence of μ_e is a much more reliable guide to the presence or absence of orbital contribution (ignoring any which is mixed into ground A or E terms from excited terms) than the simple measurement of μ_e at room temperature. Since, for any configuration, orbital contribution will occur either for octahedral or tetrahedral stereochemistries but not for both, the temperature dependence of μ_e is, therefore, the best guide in distinguishing between these stereochemistries.

SECOND ORDER ZEEMAN EFFECT

In deriving Van Vleck's formula for susceptibility it was assumed that the energy of a level in an applied magnetic field could be developed as a power series in H (equation (2)). The change in energy proportional to H is the first order Zeeman effect and is the symmetrical splitting of a level corresponding to orientation of the ion's magnetic moment with and against the field. It has been discussed extensively and gives rise to the Langevin formula for χ_M.

The change in energy due to H^2 is the second order Zeeman effect and may be regarded as a distortion of the electron distribution of the ground state by the field. This is achieved by mixing into the ground state some of the character of a higher state, causing a lowering of the ground state by an amount proportional to H^2. (This is in contrast to the first order effect which leaves the centre of gravity of the state unchanged.)

$$\frac{\partial E}{\partial H} \text{ is thus linear in } H$$

However, it has been shown (equation (4)) that $\chi_M \propto \dfrac{1}{H}\dfrac{\partial E}{\partial H}$ so that the

result of the second order Zeeman effect is to make a contribution to χ_M which is independent of temperature. It is thus distinct from the mixing in of a higher level due to spin-orbit coupling, which makes a constant contribution to the moment. (The former leads to departures from the Curie or Curie-Weiss law, the latter does not.) It is in fact the paramagnetic portion of the $N\alpha$ term introduced by Van Vleck as a correction to the Langevin formula for χ_M (equation (10), Chapter II). The separation between the interacting levels is of great importance, since not only is the magnitude of the effect dependent on it but, if it is comparable to kT, the higher level will actually be populated and a temperature dependent contribution to the susceptibility will arise. This happens in the f^6 ion, Eu^{3+}, where a ground state singlet $^7F_0$ is only about 250 cm^{-1} below the triplet $^7F_1$. At very low temperatures only the temperature independent second order Zeeman effect occurs but, at higher temperatures, the behaviour becomes complicated by population of the $^7F_1$ state.

In the f^5 ion, Sm^{3+} the first excited state $^6H_{7/2}$ is about 1 000 cm^{-1} above the ground state $^6H_{5/2}$ so that again a considerable second order Zeeman contribution occurs, though this time it is not complicated to the same extent by population of the higher state.

Where the separation of the interacting levels is much greater than kT, thermal population of the upper level does not occur and the contribution is independent of temperature. It is, therefore, frequently known as temperature independent paramagnetism (t.i.p.), or Van Vleck paramagnetism. This situation is found in transition metal ions where the interacting levels have been split by the ligand field. Because this is a comparatively large splitting, the contribution is rather small and is usually only observable at high temperatures where the normal paramagnetism is relatively small. Its value depends on the nature of the ground term.

A_1 *ground terms.* These occur only in d^5 (oct. or tet. Mn^{2+} and Fe^{3+}) and in octahedral strong field d^6 (Co^{3+}). The effect is not found with the d^5 configuration because there is no appropriate upper level available. However, the small amount of paramagnetism ($100 - 200 \times 10^{-6}$ c.g.s. units) observed in the otherwise diamagnetic, octahedral spin-paired Co^{3+} is of this type.

A_2 *ground terms.* Here the t.i.p. $= 8N\beta^2/10\,Dq$. It is observable in tetrahedral complexes of Co^{2+} where the small ligand fields lead to values as high as $400 - 600 \times 10^{-6}$ c.g.s. units.

E *ground terms.* Here the t.i.p. $= 4N\beta^2/10\,Dq$. It leads to values of approx. 60×10^{-6} c.g.s. units in octahedral compounds of Cu^{2+}.

T *ground terms.* In the preceding account of spin-orbit coupling in T terms, the second order Zeeman effect arising between components of the ground terms was actually incorporated into the values assigned to their energies. Contributions from higher levels are unlikely to be important.

In comparing experimentally obtained moments with values calculated in the above manner it is on the whole true to say that, providing magnetic dilution is adequate, agreement for A and E ground terms is good but marked differences are sometimes found for T ground terms. The most notable difference is generally that observed moments are nearer to the spin-only moment and vary less with temperature than expected. The importance of the possible operation of medium rather than weak or strong fields has already been stressed. Other factors which might lead to divergence from the predicted behaviour are as follows.

1. The effect of spin-orbit coupling in mixing higher terms into the ground term has been noted for A and E terms but not for T terms.

2. The basic assumption of crystal field theory is that the bonding between the metal and the ligands is entirely ionic. In view of the extreme nature of this assumption it is perhaps surprising that the agreement turns out to be so good. The effect of covalency is to transfer some of the "non-bonding" d electrons on to the ligands. While this does not affect the spin, it reduces any orbital angular momentum which might be present, and consequently reduces the effect of spin-orbit coupling. This is more important when dealing with T, than with A and E terms. Attempts to take account of covalency, introduce the "delocalization factor" k. This may be taken to be the fraction of time which the d electrons spend on the metal. $k = 1$ represents the extreme ionic case and is reduced as the covalency increases.

3. The assumption of perfect cubic fields, whether octahedral or tetrahedral, has been made throughout. In fact this is rarely justified. This is obvious when the ligands are not identical but, even when they are, the effect of other ions outside the immediate co-ordination sphere may be to distort the cubic symmetry. Yet another effect which can reduce the symmetry of the ligand field is known as the Jahn-Teller effect. This is expressed as: *"Any system possessing an orbitally degenerate ground term will distort itself so as to remove this orbital degeneracy"*.

The way in which a set of d orbitals splits under the influence of a ligand field was deduced by considering the repulsions exerted by the ligand electrons on the metal d electrons. A physical interpretation of the Jahn-Teller effect can be obtained by looking at this in the reverse way, i.e. by considering the repulsions exerted by the metal d electrons (which are now distributed in the separate d_ε and d_γ sets) on the ligand electrons. The configurations with a symmetrical distribution of electronic charge (octahedral d_ε^3, d_ε^6, $d_\varepsilon^3 d_\gamma^2$ and $d_\varepsilon^6 d_\gamma^2$; tetrahedral d_γ^2, $d_\gamma^2 d_\varepsilon^3$ and $d_\gamma^4 d_\varepsilon^3$) will exert symmetrical repulsions on the ligands and cause no distortion. These are the configurations with A ground terms and are unaffected by the Jahn-Teller effect.

Conversely, the remaining configurations, in which the distribution of electronic charge is asymmetrical, will cause distortion because of the greater repulsion of ligand electrons along the axes on which the higher concentrations of charge occur. These are the configurations with E and T ground terms which are split by the Jahn-Teller effect. However, since an E term is a non-magnetic doublet possessing no orbital contribution, the magnetic effect is unlikely to be important. But it is of considerable importance for T terms, since the splittings are of the same order of magnitude as those due to spin-orbit coupling. Although the results are likely to be complicated, it is a reasonable generalization that axial distortions will produce moments which are less dependent on temperature and nearer the spin-only value.*

* For references for this chapter, see groups 2, 5–8, 10 and 14 on pp. 110 and 111.

V. FURTHER TOPICS

SECOND AND THIRD ROW TRANSITION METAL COMPLEXES

The most obvious feature of the magnetism of second and third row elements, when compared to the first row, is the much greater frequency with which diamagnetism occurs and the fact that paramagnetic moments are usually much lower. Part of the reason for this is that the complexes are invariably of the spin-paired type. This is because, in the first place, the $4d$ and $5d$ orbitals are larger than the $3d$ so that the interelectronic repulsions, which tend to oppose spin pairing, are less. Also, since the nuclear charges are higher and therefore exert a stronger attraction on the ligands, a given set of ligands produces a greater splitting of the d orbitals. In fact values of 10 Dq increase by about one-third in going from the first to the second row and also from the second to the third row.

The other important reason for the low moments is that the spin-orbit coupling constants are much higher for heavier atoms, increasing by as much as a factor of three in moving from one row to the next (Table XIII). This again is due to the higher nuclear charge. As was explained in Chapter II, a higher nuclear charge has the effect of increasing the interaction between the moments associated with the spin and orbital angular momenta. Although the nucleus is appreciably screened by the filled inner shells of electrons, the penetration of these shells by the $4d$ and $5d$ electrons is sufficient to increase significantly the "effective" nuclear charge.

The equations and curves showing the dependence of μ_e on λ and T are the same as those given previously for first row ions (Table XII; Figs. 33–36). In that case, however, the magnitudes of λ were such as to give values of $\frac{kT}{|\lambda|}$ between about 0·5 and 3, leading to moments often not grossly different from $\mu_{\text{s.o.}}$. Now the values of $\frac{kT}{|\lambda|}$ are much lower, rarely rising above about

0·25, and therefore leading to moments usually a good deal lower than $\mu_{\text{s.o.}}$. The relevant curves for octahedral d^1 to d^5 ions are collected in Fig. 37 with the $\dfrac{kT}{|\lambda|}$ scale expanded appropriately for second and third row ions. These curves, and the equation from which they are derived, were originally deduced by Kotani.

Apart from the d_ε^3 configuration, with a spin-only moment independent of T (and d_ε^6 which is diamagnetic), the behaviour for very low values of $\dfrac{kT}{|\lambda|}$

TABLE XIII. *Estimated free ion spin-orbit coupling constants* (cm^{-1}) *for spin-paired octahedral complexes of second and third row transition metals*

Configuration	Ion		ζ	Ground term	λ
$4d_\varepsilon^2$	Mo	IV	850	$^3T_{1g}$	$+425$
$4d_\varepsilon^3$	Mo	III	800	$^4A_{2g}$	$+267$
$4d_\varepsilon^4$	Ru	IV	1 400	$^3T_{1g}$	-700
$4d_\varepsilon^5$	Ru	III	1 250	$^2T_{2g}$	$-1\,250$
	Rh	IV	1 700		$-1\,700$
$5d_\varepsilon^2$	W	IV	2 300	$^3T_{1g}$	$+1\,150$
	Re	V	3 700		$+1\,850$
$5d_\varepsilon^3$	W	III	1 800	$^4A_{2g}$	$+\;\;600$
	Re	IV	3 300		$+1\,100$
$5d_\varepsilon^4$	Re	III	2 500	$^3T_{1g}$	$-1\,250$
	Os	IV	4 000		$-2\,000$
	Ir	V	5 500		$-2\,750$
$5d_\varepsilon^5$	Os	III	3 000	$^2T_{2g}$	$-3\,000$
	Ir	IV	5 000		$-5\,000$

is of two types. For d_ε^2 and d_ε^5 configurations the moments vary approximately linearly with T, while for d_ε^1 and d_ε^4 they vary linearly with $\sqrt{T}$, that is to say the susceptibilities are entirely of the temperature independent type.

It is important to emphasize that these predictions apply only to magnetically dilute, octahedral compounds. Much of the reported data refer to compounds in which polynuclear formation (with consequent loss of symmetry as well as possible metal-metal interaction) or co-ordination numbers higher than six occur. The preceding theory takes no account of these possibilities. The effect of six non-equivalent ligands is probably to produce moments nearer to $\mu_{\text{s.o.}}$ which vary less with temperature than would be expected for perfect octahedral symmetry.

The compounds of di- and tetravalent Pd and Pt are invariably diamagnetic and indeed, of the later elements of these series, only for Ag(II) (d^9) has a significant number of paramagnetic compounds been examined. Having a

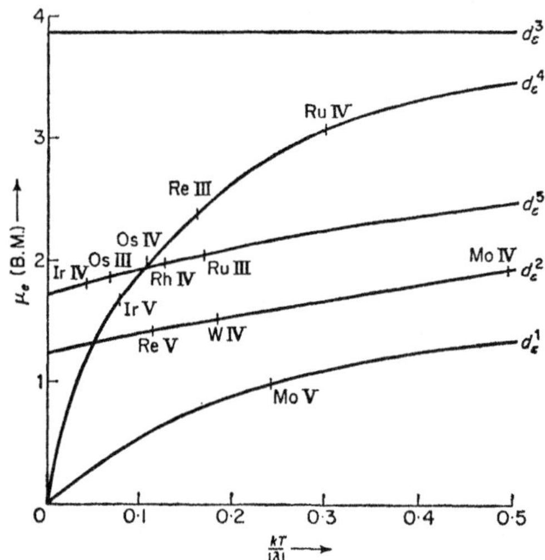

FIG. 37. The variation of magnetic moment with temperature for spin-paired octahedral complexes. The scale is chosen to cover the region occupied by second and third row transition metals. The moments for individual ions are marked for $T = 300°K$, using the free ion values of λ.

^2E_g ground term (for octahedral compounds) and $\lambda \simeq 1\,800\ \mathrm{cm}^{-1}$, moments in the region of 2 B.M. are expected. In fact they are usually appreciably nearer $\mu_{s.o.}$ but in many cases the stereochemistry is uncertain.

ANTIFERROMAGNETISM

In all the foregoing discussions it has been assumed that the individual paramagnetic ions in a compound act independently of each other. In most co-ordination compounds the shell of ligands surrounding the metal ion is sufficient to ensure this. However, the situation often arises where the individual magnetic dipoles do influence each other, leading to "co-operative" or "exchange" phenomena. This may be simply because the distance between the paramagnetic ions is small, or because the intervening atoms are capable of transmitting the magnetic interaction.

If the orbital magnetic moments of the ions are neglected, the interactions may be described by

$$\Delta E = 2J S_i S_k$$

where J is the "exchange coupling constant" (not to be confused with the quantum number J!).

If J is positive the lowest state is that with the spins aligned in the same direction and, if negative, the lowest state is that with spins paired. The former case is ferromagnetism and the latter, the much more common, antiferromagnetism.

Ferromagnetic alignment of the spin angular momenta produces susceptibilities greatly in excess of those for normal paramagnetic materials (up to 10^4 c.g.s. units) and results in the phenomena of hysteresis and marked field dependence. The S vectors are coupled parallel in groups, or "domains", and the effect of an applied magnetic field is to align these comparatively huge entities in its direction. Hysteresis arises because, on removal of the field, thermal energy is not always able to randomize this arrangement once it has been established. In addition, it is often possible to attain virtually perfect alignment of the S vectors by applying relatively moderate fields. This is known as "saturation" and it is as this situation is approached that field dependence of χ is noted, since a given increase in H is no longer able to effect the usual increase in alignment of the S vectors.

A less qualitative way of looking at this is to examine the assumptions made in Chapter II (p. 25) to calculate χ_A. Because of the possible orientations of the magnetic dipole with respect to H, a series of levels separated by $g\beta H$ are produced when a field is applied. If $g\beta H \ll kT$ then $\exp(-g\beta H/kT)$ reduces to $(1 - g\beta H/kT)$, and H finally disappears from the expression for χ_A. If, however, $g\beta H$ is comparable to kT this simplification is not possible. Higher powers of H must be retained and remain in the final expression for χ_A. This can occur, of course, with normal paramagnetic materials but, without the assistance of the exchange coupling, the field strengths required to reach this position are vastly higher than any which would be used for the measurement of χ (see Appendix).

Antiferromagnetic interaction causes a lowering of χ and μ_e and, where the interaction is direct, is the phenomenon of covalent bonding. In extreme cases such as $Fe_2(CO)_9$ and $[W_2Cl_9]^{3-}$ it results in complete spin pairing between the metal ions.

As the temperature is raised a point is reached when the thermal energy is sufficient to overcome the magnetic interaction, of whichever kind it happens to be. This is known as the Curie temperature or, in antiferromagnetism more usually, the Neel temperature. At much higher temperatures, where the interaction is negligible compared to kT, normal paramagnetic behaviour occurs and χ is independent of field and follows the Curie-Weiss law (Fig. 38).

From now on attention will be restricted to antiferromagnetism. Two types may be distinguished; intermolecular where the interactions extend throughout the crystal, and intramolecular where the interactions are confined within the molecule.

INTERMOLECULAR ANTIFERROMAGNETISM

Even in compounds in which the proportion of the paramagnetic ion is high (magnetically concentrated systems) the usual mechanism of the exchange is not direct but via intervening atoms, when it is known as "superexchange". This is particularly effective where either oxygen or fluorine provides a linear bridge between the metal ions. If, ignoring the intervening

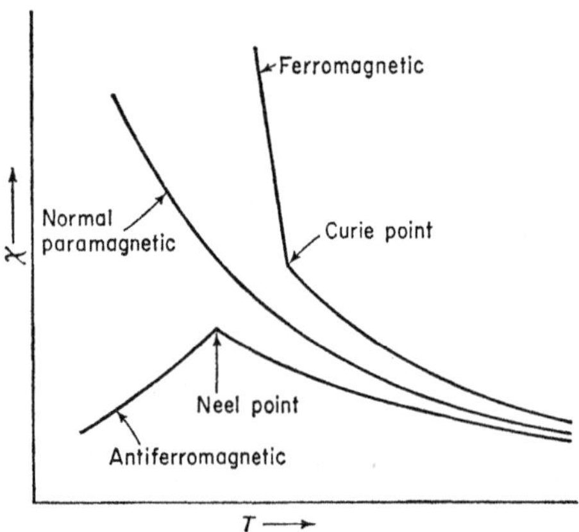

FIG. 38. The variation of susceptibility with temperature for different materials. Note that, for a ferromagnetic substance below its Curie point, χ is field dependent.

diamagnetic atoms, each paramagnetic ion is surrounded by similar ions of opposite spin, which in turn are surrounded by ions with the same spin orientation as the first, then the system is one of interpenetrating ferromagnetic lattices. Quite complicated lattice arrangements are possible, often resulting in field dependence of χ, but the main concern here is in recognizing, rather than quantitatively explaining, this effect.

Two methods are commonly used to detect this type of antiferromagnetism in co-ordination compounds. Dissolution, by surrounding the complex species with solvent molecules and thus separating the interacting ions, should destroy the interaction. The main difficulties with this method are that (a) the substances in question are frequently insoluble or are considerably changed by dissolution, and (b) the experimental errors in measuring χ, particularly if the solution is very dilute, may be greater than the effect being observed. The second method is dilution in a solid, diamagnetic, isomorphous compound. This is really the same as the solution technique and presents the difficulties of finding a suitable host material and of measuring a probably very small susceptibility.

Apart from simple oxides and halides, two well-known instances of intermolecular antiferromagnetism are the series of compounds KMF_3, where M is any of several divalent first row transition metal ions, and K_2MX_6 where M is a second or third row transition metal ion and X is a halogen.

Intramolecular Antiferromagnetism

In polynuclear complexes, when the interacting ions are present within a single molecule, isomorphous dilution and dissolution have no effect on the interaction—unless, of course, the complex is broken up by the solvent. The mathematical treatment of bi- and tri-nuclear complexes has been developed (reference 7).

Binuclear compounds

First, the simplifying assumptions are made that the interacting ions are identical and that their ground terms are effectively S terms (i.e. they are A or E and have no first order orbital contribution).

If each ion has a spin angular momentum specified by the quantum number S, then the molecule has a total angular momentum specified by the quantum number S'. This can have the values 0 to $2S$ in integral steps: a total of $(2S+1)$ values. Each of these values corresponds to a particular energy level. A given level has a multiplicity of $2S'+1$ and its energy is $J[S'(S'+1)]$ above the ground level, where J is the exchange coupling constant. Not infrequently the value of J is defined in such a way as to be twice that used here and the energy of a level is then $\frac{1}{2}J[S'(S'+1)]$. In comparing numerical values from different sources this point must obviously be checked.

The minus sign, necessarily associated with J if the interaction is antiferromagnetic, is taken into account in the order of the levels. $S' = 0$ defines the ground level and $S' = 2S$ defines the highest. (If the interaction were ferromagnetic the ground level would be that with parallel spins, i.e. $S' = 2S$.) Figure 39 illustrates the separation of the levels.

The problem now is exactly the same as with multiplet widths comparable to kT, and the susceptibility of *the molecule* is obtained by applying Van Vleck's equation (Chapter IV):

$$\chi_M = \frac{N \sum \left(\dfrac{E_i{}^2{}_{(1)}}{kT} - 2E_{i\,(2)} \right) \exp \left(-\dfrac{E_{i\,(0)}}{kT} \right)}{\sum \exp \left(-\dfrac{E_{i\,(0)}}{kT} \right)} \qquad (1)$$

The first order Zeeman effect is to split each level into its $2S'+1$ component levels, ranging in energy from $-g\beta HS'$ to $+g\beta HS'$

$$\therefore \quad \frac{E_i{}^2{}_{(1)}}{kT} \text{ becomes } \frac{g^2\beta^2}{kT} [(S')^2 + (S'+1)^2 + \ldots 0 \ldots (-S')^2]$$

$$= \frac{g^2\beta^2}{kT} \cdot \frac{S'(S'+1)(2S'+1)}{3}$$

The denominator of equation (1) must include the factor $(2S'+1)$ since each component of a degenerate set of levels must be counted separately. If the second order Zeeman term is incorporated into the $N\alpha$ term, equation (1) reduces to

$$\chi_M = \frac{Ng^2\beta^2}{3kT} \frac{\sum S'(S'+1)(2S'+1)\exp\left(-\frac{E_{i(0)}}{kT}\right)}{\sum (2S'+1)\exp\left(-\frac{E_{i(0)}}{kT}\right)} + N\alpha \qquad (2)$$

The way in which equation (2) is applied can be seen by taking a specific example, say that of a binuclear molecule with identical paramagnetic ions

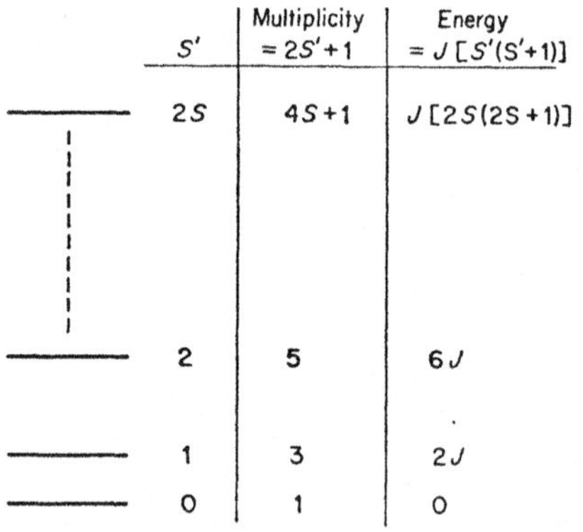

S'	Multiplicity $= 2S'+1$	Energy $= J[S'(S'+1)]$
$2S$	$4S+1$	$J[2S(2S+1)]$
2	5	$6J$
1	3	$2J$
0	1	0

FIG. 39. The multiplicities and energies produced by exchange interaction in a binuclear complex.

with $S = 3/2$. Four levels occur corresponding to $S' = 0$, 1, 2 and 3 with energies of 0, $2J$, $6J$ and $12J$.

$$\therefore \quad \chi_M =$$

$$= \frac{Ng^2\beta^2}{3kT} \frac{\left[0+1.2.3\exp\left(-\frac{2J}{kT}\right)+2.3.5\exp\left(-\frac{6J}{kT}\right)+3.4.7\exp\left(-\frac{12J}{kT}\right)\right]}{\left[\exp(0)+3\exp\left(-\frac{2J}{kT}\right)+5\exp\left(-\frac{6J}{kT}\right)+7\exp\left(-\frac{12J}{kT}\right)\right]} + N\alpha$$

multiplying throughout by x^{12} where $x = \exp(J/kT)$ and substituting

$$\chi_A = \tfrac{1}{2}\chi_M:$$

$$\chi_A = \frac{Ng^2\beta^2}{3kT} \cdot \frac{[42+15x^6+3x^{10}]}{[7+5x^6+3x^{10}+x^{12}]} + N\alpha$$

The equation for other values of S are derived in the same manner and are collected below.

For $\quad S = \tfrac{1}{2} \qquad \chi_A = \dfrac{3K}{T}\left[\dfrac{1}{3+x^2}\right] + N\alpha$ (3)

$$S = 1 \qquad \chi_A = \dfrac{3K}{T}\left[\dfrac{5+x^4}{5+3x^4+x^6}\right] + N\alpha \tag{4}$$

$$S = \tfrac{3}{2} \qquad \chi_A = \dfrac{3K}{T}\left[\dfrac{14+5x^6+x^{10}}{7+5x^6+3x^{10}+x^{12}}\right] + N\alpha \tag{5}$$

$$S = 2 \qquad \chi_A = \dfrac{6K}{T}\left[\dfrac{30+14x^8+5x^{14}+x^{18}}{9+7x^8+5x^{14}+3x^{18}+x^{20}}\right] + N\alpha \tag{6}$$

$$S = \tfrac{5}{2} \qquad \chi_A = \dfrac{3K}{T}\left[\dfrac{55+30x^{10}+14x^{18}+5x^{24}+x^{28}}{11+9x^{10}+7x^{18}+5x^{24}+3x^{28}+x^{30}}\right] + N\alpha \tag{7}$$

where $x = \exp(J/kT)$ and $K = Ng^2\beta^2/3k = 0\cdot1251\,g^2$.

If $N\alpha$ is neglected, then substitution of $J = 0\,(\therefore\ x = 1)$ and $g = 2$, followed by insertion of the values of χ_A in Langevin's formula leads to the spin-only moments of $\sqrt{3}$, $\sqrt{8}$, $\sqrt{15}$, $\sqrt{24}$ and $\sqrt{35}$ B.M.

The units of J should be noted. As an energy term it is conventionally, if rather loosely, measured in cm^{-1} units, when

$$\frac{J}{kT} = \frac{J\,(\text{in cm}^{-1})}{0\cdot69503 \times T\,(\text{in degrees})}.$$

It is also frequently measured in "degrees" by incorporating k, when

$$\frac{J}{kT} = \frac{J\,(\text{in "degrees"})}{T\,(\text{in degrees})}$$

The conversion factor is therefore 1 degree $\equiv 0\cdot69503\ \mathrm{cm}^{-1}$ (see Appendix).

In Fig. 40, $\chi_A{}^{-1}$ is plotted against T, for $S = \tfrac{1}{2}$ ions with increasing values of J. This illustrates the way in which increased interaction leads to lower susceptibilities (and hence lower moments), higher Neel points and greater curvature (i.e. departure from the Curie-Weiss law) above the Neel points. In the extreme, diamagnetism results.

If the binuclear molecule is composed of different paramagnetic ions (i and k) the appropriate equations, though naturally more complicated, can be derived in precisely the same way using values of S' from $S_i - S_k$ to $S_i + S_k$. The importance of the assumption of an effectively S ground term must be emphasized. If the moment in the absence of interaction is exactly $\mu_{\mathrm{s.o.}}$ then $g = 2$, but g will depart from this value depending on the orbital contribution. If this is achieved by the mixing of higher terms into the ground term

(A or E ground term), then g should be independent of temperature and the formulae still hold with $g \neq 2$.

The further tacit assumption has been made that metal–metal interactions

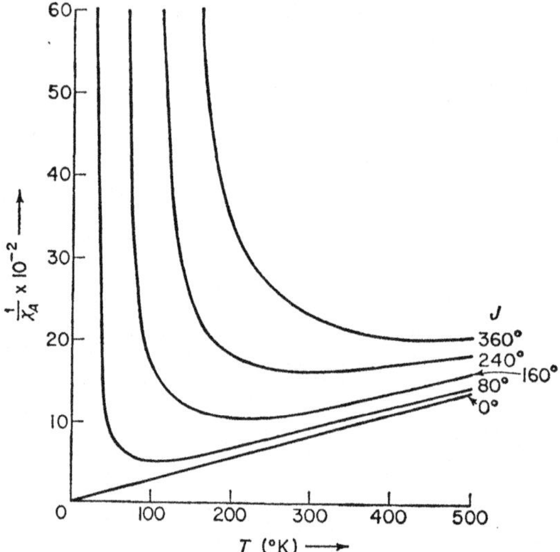

FIG. 40. The plots of $1/\chi_A$ against T, showing the effects of increasing exchange interaction, for binuclear molecules where $S_i = S_k = \frac{1}{2}$. For simplicity the $N\alpha$ term has been neglected and g is assumed to be exactly 2.

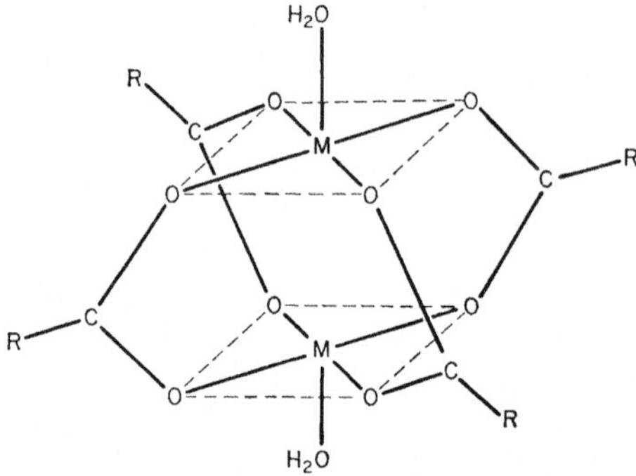

FIG. 41. The structure of the binuclear acetates of Cu^{2+} and Cr^{2+}. In this case R is CH_3.

are confined within the molecule and do not take place throughout the lattice. Ideally this should be checked by measurements on solutions.

Well-known examples of binuclear compounds are the acetates of Cu^{2+} and Cr^{2+} which occur in the form $M_2(CH_3.COO)_4$ where the two metal ions are held together by four acetate bridges (Fig. 41). In the hydrates water molecules are positioned above and below the molecules as indicated.

The precise mechanism of the interaction is not known with certainty but it would appear that direct exchange is possible by sideways overlap of the $d_{x^2-y^2}$ orbitals and, in the case of Cr^{2+}, also by overlap of the $d_{xz}d_{yz}$ and d_{z^2} orbitals of the two metal ions. Equation (3) gives a good fit with the experimental values for copper acetate, anhydrous and hydrated, using values of $g = 2 \cdot 17$ and $2 \cdot 13$, $J = 150$ and 142 cm^{-1} respectively and $N\alpha = 60 \times 10^{-6}$ in both cases. Chromous acetate is complicated by the fact that it is invariably contaminated by traces of chromic ion. J cannot be estimated exactly but must be of the order of 400 cm^{-1}, producing only very feeble paramagnetism.

Binuclear compounds in which the metal ions are linked by 1, 2 or 3 ligand bridges provide many other examples of this type of interaction.

Trinuclear compounds

Three interactions are possible and may occur in a variety of ways. The simplest, which has been examined by Kambe, is that in which the three metal ions are identical and are arranged at the corners of an equilateral triangle, producing three equal interactions. The total spin angular momentum vector is now the vectorial sum of three individual spin vectors, so the quantum number S' can have integral values from 0 or $\frac{1}{2}$ up to $3S$. An important difference compared to binuclear molecules is that the multiplicity of each of these levels is no longer simply $(2S'+1)$, since a particular value of S' may arise in a number of ways. This is best understood by taking a specific example.

$$\text{Let } S_1 = S_2 = S_3 = \tfrac{3}{2} \text{ (e.g. trinuclear } Cr^{3+})$$

First of all, two of the vectors are considered to couple in the usual way, the resultant being associated with a quantum number (S_1+S_2) with values of 0, 1, 2 or 3. This vector now couples with the remaining spin vector to give the total angular momentum vector with which S' is associated. S' may take the following values:

$$
\begin{array}{ll}
\text{for } S_1+S_2 = 0 & S' = \tfrac{3}{2} \\
\quad\quad S_1+S_2 = 1 & S' = \tfrac{1}{2}, \tfrac{3}{2}, \tfrac{5}{2} \\
\quad\quad S_1+S_2 = 2 & S' = \tfrac{1}{2}, \tfrac{3}{2}, \tfrac{5}{2}, \tfrac{7}{2} \\
\quad\quad S_1+S_2 = 3 & S' = \tfrac{3}{2}, \tfrac{5}{2}, \tfrac{7}{2}, \tfrac{9}{2}
\end{array}
$$

As a result the factors 2, 4, 3, 2 and 1 must be introduced into the multiplicities of the levels corresponding to $S' = \tfrac{1}{2}, \tfrac{3}{2}, \tfrac{5}{2}, \tfrac{7}{2}$ and $\tfrac{9}{2}$ in that order. The energy of a set of levels is $J[S'(S'+1)]$ as illustrated in Fig. 42.

Again the minus sign associated with J, because the interaction is antiferromagnetic, has been taken into account in the order in which the levels occur. Since the lowest in this case corresponds to $S' = \tfrac{1}{2}$ rather than 0, a finite energy is ascribed to it. However only differences in energy have any effect in the formula for χ_A and any arbitrary energy could be ascribed to the lowest level. Remembering the factors just deduced for the multiplicities, and substituting the appropriate values for S' and $E_{i(0)}$ in equation (2),

$$\chi_M = \frac{Ng^2\beta^2}{3kT} \cdot$$

$$\frac{\left[\frac{9}{2}\cdot\frac{11}{2}\cdot 10\exp\left(-\frac{99}{4}\frac{J}{kT}\right)+2\cdot\frac{7}{2}\cdot\frac{9}{2}\cdot 8\exp\left(-\frac{63}{4}\frac{J}{kT}\right)+3\cdot\frac{5}{2}\cdot\frac{7}{2}\cdot 6\exp\left(-\frac{35}{4}\frac{J}{kT}\right)\right.}{10\exp\left(-\frac{99}{4}\frac{J}{kT}\right)+2.8\exp\left(-\frac{63}{4}\frac{J}{kT}\right)+3.6\exp\left(-\frac{35}{4}\frac{J}{kT}\right)}$$

$$\frac{\left.+4\cdot\frac{3}{2}\cdot\frac{5}{2}\cdot 4\exp\left(-\frac{15}{4}\frac{J}{kT}\right)+2\cdot\frac{1}{2}\cdot\frac{3}{2}\cdot 2\exp\left(-\frac{3}{4}\frac{J}{kT}\right)\right]}{+4.4\exp\left(-\frac{15}{4}\frac{J}{kT}\right)+2.2\exp\left(-\frac{3}{4}\frac{J}{kT}\right)\right]}+N\alpha$$

multiplying throughout by $x^{99/4}$ where $x = \exp(J/kT)$ and substituting $\chi_A = \frac{1}{3}\chi_M$:

$$\chi_A = \frac{K}{4T}\left[\frac{165+168x^9+105x^{16}+40x^{21}+2x^{24}}{5+8x^9+9x^{16}+8x^{21}+2x^{24}}\right]+N\alpha \qquad (8)$$

where, as before,

$$K = \frac{Ng^2\beta^2}{3k} = 0.1251\ g^2$$

The other situation of interest is that in which $S_1 = S_2 = S_3 = \frac{5}{2}$ (e.g. trinuclear Fe^{3+}) the equation for which turns out to be

$$\chi_A =$$
$$\frac{K}{4T}\left[\frac{340+455x^{15}+429x^{28}+330x^{39}+210x^{48}+105x^{55}+20x^{60}+x^{63}}{4+7x^{15}+9x^{28}+10x^{39}+10x^{48}+9x^{55}+4x^{60}+x^{63}}\right]+N\alpha$$
$$(9)$$

Neglecting $N\alpha$ and substituting in equations (8) and (9), $J = 0$ ($\therefore\ x = 1$) and $g = 2$ and inserting the values of χ_A in Langevin's formula leads to the spin only moments of $\sqrt{15}$ and $\sqrt{35}$ B.M. respectively.

The equations for other trinuclear molecules, whether of other spins or of mixed paramagnetic ions, can be derived similarly.

An interesting consequence of having three identical interactions within the molecule is that, if S is half integral, the lowest level must correspond to $S' = \frac{1}{2}$ and the paramagnetism can never be completely destroyed.

The best known examples of trinuclear molecules, in which the paramagnetic ions form an equilateral triangle, are several carboxylates of Cr^{3+} and

Fe^{3+}. These have been formulated as $[M_3(\text{carboxylate})_6(\text{OH})_2]^+$ but the structure of the chromic acetate complex has been shown by X-ray crystallographic methods to be that illustrated in Fig. 43. It therefore seems likely

S'	Total multiplicity	Energy, $E_{i(0)}$ $= J[S'(S'+1]$
$\frac{9}{2}$	1 x 10	$\frac{99}{4}J$
$\frac{7}{2}$	2 x 8	$\frac{63}{4}J$
$\frac{5}{2}$	3 x 6	$\frac{35}{4}J$
$\frac{3}{2}$	4 x 4	$\frac{15}{4}J$
$\frac{1}{2}$	2 x 2	$\frac{3}{4}J$

FIG. 42. The multiplicities and energies produced by three equal exchange interactions in a trinuclear molecule composed of three identical metal ions, each with $S = 3/2$.

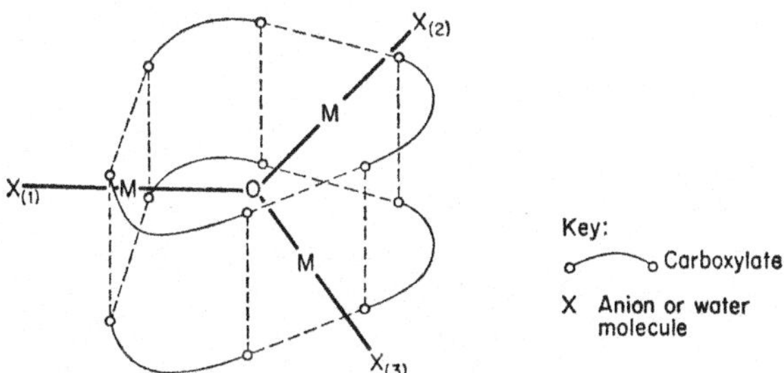

FIG. 43. The structure of trivalent metal carboxylates $[M_3(\text{carboxylate})_6O]^+$.

that most of these compounds should be formulated similarly as $[M_3(\text{carboxylate})_6O]^+$.

The problem of compounds in which identical ions are arranged in linear chains of up to ten members has been examined by Figgis, on the assumption that exchange occurs between adjacent members only. For a three membered chain the expressions for $S = \frac{3}{2}$ and $\frac{5}{2}$ are again equations (8) and (9). The only difference is that for the linear system there are only two interactions instead of three. Consequently, to produce the same effect, the value of J (linear) $= \frac{3}{2} J$ (triangular).

SPIN FREE–SPIN PAIRED EQUILIBRIA

Ligand fields are able to induce spin-pairing of electrons in certain con-figurations by virtue either of their magnitude, if octahedral, or of their asymmetry. For the configurations d^4 to d^7 a sufficiently large octahedral field causes spin-pairing within the lower d_ε orbitals. Alternatively, as the Tanabe-Sugano diagrams show, the ligand fields may be said to cause a change of ground term such that the new ground term is of lower multiplicity. This is not possible for other configurations but, for d^8, a tetragonal distor-tion which ultimately results in a square planar stereochemistry may separate the $d_{x^2-y^2}$ and d_{xy} orbitals sufficiently to force spin-pairing. Again this may be interpreted as the lowering of a singlet below the triplet term.

Near the cross-over point of the relevant terms the difference in energy of the spin free and spin paired configurations is comparable to the thermal energy, kT. Consequently, the susceptibility is given by the population-weighted average of the susceptibilities of the two configurations. Actually in the case of octahedral d^4 and d^7 this is an oversimplification since, for these ions, spin-orbit coupling mixes the two terms and leads to very compli-cated expressions for the susceptibilities. However, for the other two octa-hedral configurations, d^5 and d^6, the general formula of equation (10) applies:

$$\chi_A = \frac{(2S_1+1)\chi_1+(2S_2+1)\chi_2\exp\left(-\dfrac{E}{kT}\right)}{(2S_1+1)+(2S_2+1)\exp\left(-\dfrac{E}{kT}\right)} \tag{10}$$

where χ and S give the susceptibilities and total spin of the terms represented by the subscripts 1 and 2. E is the excess energy of term 2 above 1. Specifi-cally this refers to the d^5 ions Mn^{2+} and Fe^{3+}, where the two terms are $^6A_{1g}$ and $^2T_{2g}$, and the d^6 ions Fe^{2+} and Co^{3+}, where the two terms are ^5E_g and $^1A_{1g}$. The expressions for χ_1 and χ_2 (which are themselves temperature dependent) can be obtained from Chapter IV. In applying equation (10) it must be realized that either term may be lowest depending on which side of the cross-over point the particular system lies. One of the few definitely established examples of this type of behaviour is the ferric complex Fe(di-ethyl dithiocarbamate)$_3$.

Spin free–spin paired equilibria are rather better documented for d^8, Ni^{2+}. The phenomenon can arise from a square planar-octahedral equilibrium, as in the case of Ni(N-methylsalicylaldoxime)$_2$ dissolved in pyridine, or from a square planar-tetrahedral equilibrium, as in the case of $(Bu^nPh_2P)_2NiX_2$ dissolved in benzene. It is important to distinguish between an equilibrium of this sort which involves an actual change in stereochemistry between two distinct forms and one in which the stereochemistry is rigid but in which singlet or triplet levels lie close together. The above compounds of nickel(II) in fact occur in two distinct forms in the solid. It is only in their solutions or in melts, where stereochemical changes (particularly due to the influence or approach of solvent molecules) occur easily, that the equilibrium is reversibly temperature dependent.

In cases where the spin-paired form is of lower energy than the spin-free then, remembering that for each form $\chi_A = \dfrac{Ng^2\beta^2}{3kT} \cdot S(S+1)$, equation (10) becomes

$$\chi_A = \frac{0 + 3 \cdot \dfrac{Ng^2\beta^2}{3kT} \cdot 2\exp\left(-\dfrac{E}{kT}\right)}{1 + 3\exp\left(-\dfrac{E}{kT}\right)}$$

multipiying throughout by x where $x = \exp(E/kT)$,

$$\chi_A = \frac{Ng^2\beta^2}{3kT} \cdot \left[\frac{6}{3+x}\right] \tag{11}$$

Since they both describe singlet–triplet transitions, equation (11) is similar to equation (3) obtained for binuclear complexes of Cu(II). The difference, of course, is that in the latter case there are two atoms involved and so the atomic susceptibility is half that of equation (11), and $2J$ occurs instead of E.*

* For references to this chapter, see groups 2, 5–8, 10 and 14 on pp. 111 and 112.

VI. EXPERIMENTAL

When an experimentalist makes a measurement the interpretation of the result is a logical extension of the work. The object of this chapter is therefore to provide an account of the more important experimental methods and then to show briefly the way in which the problems of interpretation may be tackled.

MEASUREMENTS

Many different methods have been used to measure the magnetic properties of materials in solid, liquid and gaseous states. Most of the measurements performed by inorganic chemists have been on paramagnetic solids and to a lesser extent liquids, and attention will be confined to the methods most widely used for these purposes.

GOUY METHOD

This is the simplest of all and the most common. It has the advantage that the apparatus is simple and robust, and can be used to measure a wide range of susceptibilities. It consists essentially of suspending a uniform rod of the specimen in a non-homogeneous magnetic field, and measuring, by a conventional weighing technique, the force exerted on it.

If the field gradient over an element of volume, ∂v, of the specimen is $\partial H/\partial l$, then the force, ∂F, exerted is

$$\partial F = (\kappa_1 - \kappa_2)\,\partial v\, H \frac{\partial H}{\partial l} \tag{1}$$

where κ_1 and κ_2 are the susceptibilities per unit volume of the specimen and the displaced medium respectively.

Integration over the whole length of the specimen from $H = H_1$ at zero length to $H = H_0$ at length $= l$ gives

$$F = \frac{(\kappa_1 - \kappa_2)\, v\, (H_1{}^2 - H_0{}^2)}{2l} \qquad (2)$$

If F, in dynes, is replaced by wg, where w is in grams and $(H_1{}^2 - H_0{}^2)$ by H^2, then

$$\kappa_1 = \kappa_2 + \frac{2lgw}{vH^2}$$

but

$$\kappa = \chi\rho = \frac{\chi W}{v}$$

where W is the total weight of the specimen

$$\therefore \quad \chi = \frac{\kappa_2 v + \left(\dfrac{2gl}{H^2}\right) w}{W} = \frac{A + Bw}{W} \qquad (3)$$

Figure 44 shows, diagrammatically, the system used for measurements at room temperature only. The specimen is maintained in a suitable shape by placing it in a flat-bottomed glass tube, suspended from the pan of a semi-micro balance (sensitive to 2×10^{-5} g). The whole suspension is enclosed in glass to exclude draughts, and a thermometer is positioned to measure the temperature.

Probably the most convenient form of suspension is a fine non-ferrous chain which is flexible but cannot be stretched. If the glass tube is slightly lipped, an aluminium collar with small holes drilled diametrically opposite each other can be fitted with a springy metal stirrup which may be attached to a hook on the suspension chain.

Ideally the specimen and magnet should be such that the base of the specimen is in uniform maximum field, while the top is in zero field. In this way, not only is the whole of the field utilized but slight changes in vertical position, occasioned by movement of the balance beam during weighing, have no effect. This unfortunately requires specimens which are inconveniently long, or fields which are too small, or magnets which are too large and expensive, and a compromise is therefore necessary. With a magnet pole gap of between 1·5 and 2 cm, and a pole face diameter of 2 or 3 cm, small permanent or electromagnets may be obtained producing fields of about 5 000 oersteds. If the balance is fitted with a rider adjustment on its beam, the movement of the sample during weighing may be virtually eliminated, and a specimen length of 4 or 5 cm is then adequate.

Powdered solids are packed into the glass specimen tube a little at a time, the tube being tapped on a hard surface several times between each addition, until the required length is obtained. If the solid has to be ground initially, care must be taken to ensure that a uniform particle size is attained. If it is already of a fairly fine crystalline form it is usually best to leave well alone! With a tube of, say, 3 mm internal diameter, 0·5 g of solid may be needed, though this naturally varies considerably. In the case of solutions it is generally necessary to use a larger amount, and tubes of 1 cm diameter are better.

The procedure for measuring the pull on the specimen is simply to suspend

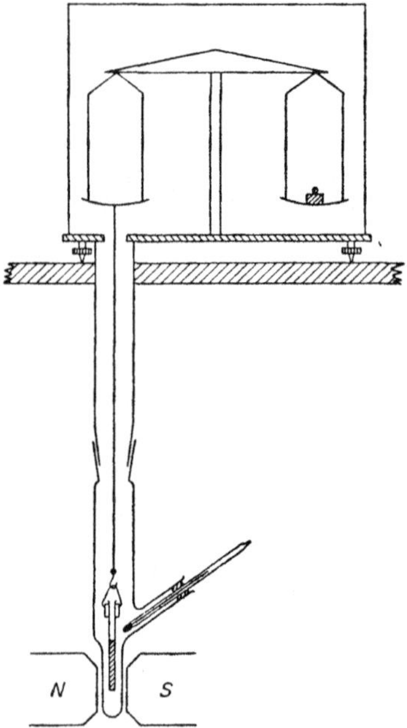

FIG. 44. Diagrammatic representation of a single temperature Gouy balance.

it, weigh it without the field and then again with the field applied. With an electromagnet this merely involves switching the current off and on, but a permanent magnet must be moved away and towards the sample. This may be achieved either by mounting it on a trolley fitted with a screw mechanism, or by hinging it to the floor and swinging it away. In either case a suitable stop can be fitted to provide a reproducible "on" position.

To be able to allow for the diamagnetism of the glass tube it is necessary first to weigh the empty tube with the field on and off, the difference being w_t (when using an electromagnet a series of values should be obtained for different fields). When, subsequently, the tube is filled with the sample and weighed with and without the field, w_{t+s} is obtained. w is then the difference $w_{t+s} - w_t$. Since glass is diamagnetic w_t is a negative quantity and, for paramagnetic samples, $w > w_{t+s}$.

W is obtained as the difference between the weights of the filled and empty tube, without field.

Though a permanent magnet has the advantages of cheapness and simplicity, an electromagnet is undoubtedly more useful. The range of paramagnetic susceptibilities is such that a single field strength is either too low to produce an accurately measurable pull for the low susceptibilities, or too high to prevent lateral movement of the specimen (causing the tube to "stick" to the pole face) for the high susceptibilities. The latter difficulty can to some extent be countered by using heavier tubes with smaller bores, but this considerably increases the difficulty of packing uniformly. A further advantage of an electromagnet is that χ can be measured at more than one field strength. If χ is found to be independent of H it can be taken that no ferromagnetic impurities are present. If, on the other hand, χ is dependent on H an attempt may be made to correct for this. At high field strengths a ferromagnetic material approaches "saturation" and its susceptibility becomes smaller. By plotting χ against $1/H$ and extrapolating to infinite field, it might he hoped that the effect of the impurity would be eliminated and that the limiting value of χ so obtained would be that of the uncontaminated sample. With the relatively high fields which are used with this method it is indeed usually possible to effect a practically linear extrapolation. However, since the specimen is in an inhomogeneous field, varying from a high to a very low value, it is evident that the interpretation is far from simple and the value of χ so obtained would only be acceptable on the basis that "any correction is better than nothing"!

Calibration

The constants A and B in equation (3) must be obtained by calibration before measurements can be made. A is the product of κ_2 (for air $\simeq +0.029 \times 10^{-6}$ at room temperature), and v is obtained from the weight of water needed to fill the tube to the required level (marked on the glass). For subsequent measurements on solids, a correction for the volume of the meniscus ($\simeq \pi r^3/3$) may be necessary, but can be omitted if the subsequent measurements are to be on solutions. B is found by measuring w and W for a substance of known χ and, if an electromagnet is used, values corresponding to each field must be obtained.

Two very good solid calibrants are $HgCo(CNS)_4$ and $Ni(en)_3(S_2O_3)$. They are easily prepared pure, do not decompose or absorb moisture, and pack well. Their susceptibilities at $20°$ C are 16.44×10^{-6} and 11.03×10^{-6} c.g.s. units, decreasing by 0.05×10^{-6} and 0.04×10^{-6} per degree temperature rise respectively, near room temperature. The cobalt compound, besides having the higher susceptibility, also packs rather densely and is suitable for calibrating low fields, while the nickel compound with lower susceptibility and density is suitable for higher fields.

D

For solution measurements a liquid calibrant, eliminating any possible packing errors, is to be preferred. The susceptibility of pure water at 20° C is -0.720×10^{-6} c.g.s. ($|\chi|$ increases by approx. 0.0009×10^{-6} per degree rise near 20° C) but oxygen should be excluded, otherwise this value may be appreciably reduced. Solutions of nickel chloride may be used if a higher susceptibility is required. For solutions of about 30% by weight of $NiCl_2$ χ is given by

$$\chi = \left[\frac{10\,030\,p}{T} - 0.720(1-p)\right] \times 10^{-6}$$

where p is the weight fraction of $NiCl_2$ in the solution, and T is measured in °K. The disadvantage of this is that p must be determined by chemical analysis which may introduce a significant error. This can be avoided by using solutions of Cs_2CoCl_4 which are made up directly from known weights of solid without analysis of the solution. χ is given by

$$\chi = \left[\frac{6\,867\,p}{T+18} - 0.720(1-p)\right] \times 10^{-6}$$

where p is now the weight fraction of Cs_2CoCl_4 in the solution.

If there is any doubt about the length of the specimen required, a check can be made by measuring w_{t+s} with the specimen up to the determined mark and using the balance rider adjustment deliberately to obtain readings at different points on the illuminated scale of the balance. If the accompanying changes in the position of the sample cause no sensible change in the pull, then the length is adequate.

Sources of error

The main error in the Gouy method for solids arises from the inhomogeneous packing of the sample. This can be reduced by repeating measurements on repacked samples until relatively constant values of χ are obtained. Agreement within 1% is to be considered good and, in some cases, much higher discrepancies are unavoidable.

For substances with such low susceptibilities that w is subject to appreciable balance errors, some improvement is possible by increasing the diameter of the tube. However, it must be remembered that w is a small difference between two comparatively large quantities, and increasing W in this way eventually introduces errors due to the instability of the suspended system. Lateral oscillation of the sample may cause a noticeable change in the balance reading. With small samples this is easily prevented, but with dilute solutions a total load of over 50 g is not uncommon and then even minute lateral movements have a serious effect. The sample must be left suspended for a considerable time before taking a reading, and the glass column surrounding it should be well lagged to reduce convection currents produced by external draughts.

With dilute solutions it is better to use a much larger magnet so that, with a pole gap of 3 cm and pole tips tapered from about 10 cm down to 6 cm diameter, fields of 10 000 oersteds are possible. Although longer specimens (10 cm) are necessary this is rarely prohibitive. In order to avoid the need to apply tube and solvent corrections, double length tubes as shown in Fig. 45

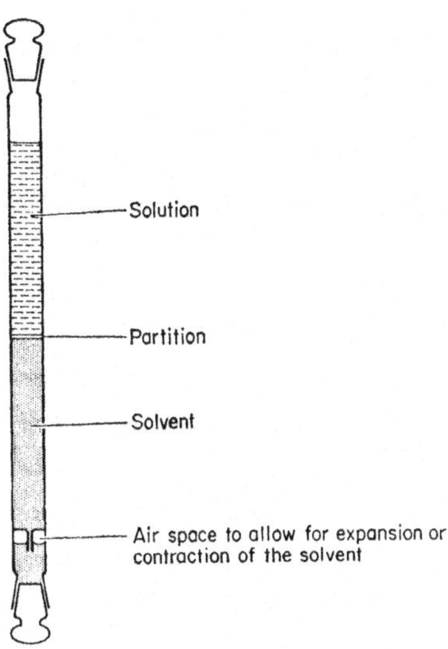

FIG. 45. Double length Gouy tube used to eliminate tube and solvent corrections.

are sometimes used. The solution is placed in the upper part and the pure solvent in the lower part, which has a small expansion space. The diaphragm is central with respect to the field and, if the field is symmetrical and the bore of the tube exactly uniform, the difference in susceptibility between solution and solvent is measured.

Variation of temperature

Measurements at low temperatures are generally of greater interest to co-ordination chemists, and most attention has therefore been given to ways of cooling below rather than heating above room temperature. To perform measurements over a range of temperatures the apparatus must be modified considerably, though the principle remains the same. The specimen has to be surrounded by a cryostat in order to control the temperature, and the bulky nature of this necessitates the use of pole gaps of up to 6 cm with a pole face diameter of perhaps 10 cm. To maintain a reasonable field with these dimensions an electromagnet is almost essential. It is advisable, too, to replace the glass sample tube with perspex or quartz, since glass frequently contains paramagnetic or traces of ferromagnetic impurities, causing the tube correction to vary appreciably with temperature and necessitating a detailed calibration of the tube over the whole range of temperature.

The simplest form of temperature control, and one which only requires a 3 cm pole gap, is provided by using constant temperature baths. The jacket surrounding the specimen tube is immersed in an appropriately shaped dewar flask containing a suitable refrigerant (Fig. 46). The availability of refrigerants limits the number of obtainable temperatures and, except when boiling liquids are used, it is usually necessary to dismantle the cryostat in order to replace the coolant.

A more satisfactory arrangement is to pass nitrogen vapour, from a boiling reservoir, around the jacket. By electrically controlling the rate of boiling, and

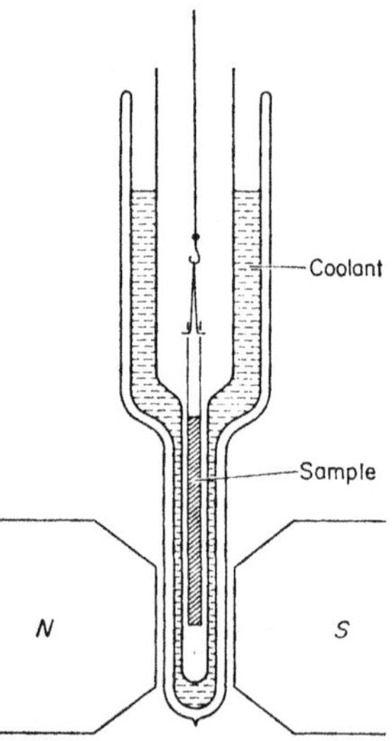

FIG. 46. A simple temperature control system for a Gouy balance.

hence the flow of vapour, the temperature of the sample is varied. Temperature gradients are reduced if the sample is surrounded by a metal of good thermal conductivity and the temperature may be measured if a thermocouple is inserted near the sample (Fig. 47). Difficulty is, however, experienced in reducing the temperature below about 100° K.

Though it requires wide pole gaps, the most generally satisfactory control is obtained by balancing the heat loss from a chamber surrounding the sample, to a reservoir of liquid nitrogen, by an electrical heater (Fig. 48). The chamber can be made of copper or aluminium, but brass has the advantage of being more easily machined. It should have an internal diameter of not less than 1·3 cm and be provided with a four start thread on the outside to accommodate the heater and platinum resistance control, wound non-inductively. The heating current is automatically controlled by a bridge

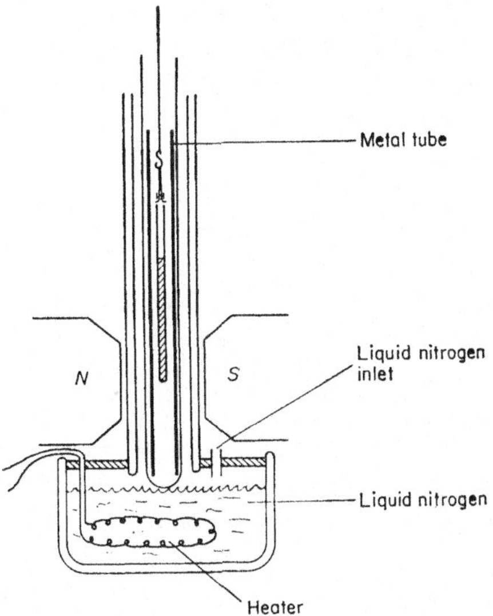

FIG. 47. A cryostat depending for its operation on the variation, electrically, of the rate of boiling of a reservoir of liquid nitrogen.

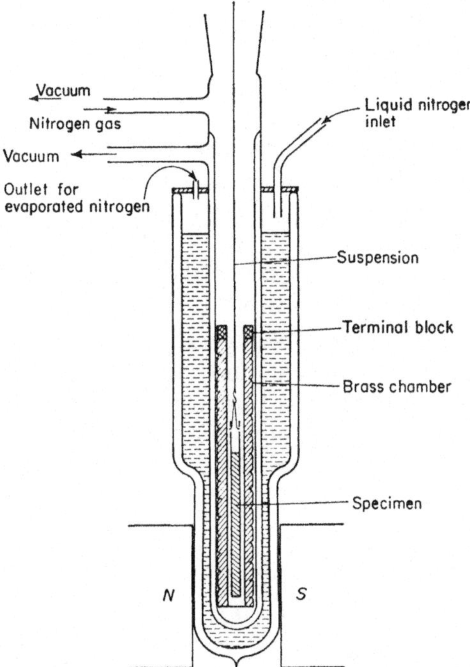

FIG. 48. A cryostat allowing automatic temperature control for a Gouy balance.

arrangement, one arm of which is the platinum resistance and the other arm a variable resistance. The "out of balance" current of this bridge is made to switch the heating current on and off or alternatively to add an extra increment to the heating current. A continuous range of temperatures can be obtained according to the setting of the variable resistance.

In order to avoid using large heating currents which would not only cause excessive evaporation of liquid nitrogen but would also produce convection currents around the sample, the inside jacket must be double-walled. The intervening space can then be evacuated to reduce heat losses at the higher temperatures. Although an adjustable leak may be fitted to the vacuum system, adequate and simpler control is possible merely by releasing the vacuum completely at temperatures below about 150° K.

Because of the differing thermal expansion coefficients of the materials employed, the platinum resistance cannot be used to measure the temperature though this is possible if the platinum is replaced by copper (this presents difficulties in winding, since a much greater length or smaller cross section is required). Instead a thermocouple, inserted inside the chamber, can be used to measure the temperature.

The level of liquid nitrogen in the outer dewar must be maintained above the top of the chamber, and this too can be done automatically. If the system is allowed to equilibrate for 1 hr before a measurement is taken, convection currents around the sample have no noticeable effect. Indeed a much shorter time is frequently sufficient.

With specimens 10 cm long one of the chief problems of temperature control is to eliminate, or at least minimize, temperature gradients along this length. In this type of cryostat the best results are achieved by siting the specimen relatively near to the bottom of the chamber.

Full details of construction and useful electrical circuits are available (references 8 and 11).

Any measurement near, or below, the boiling point of oxygen ($\simeq 90°$ K) must be performed in the absence of air. Nitrogen provides a convenient substitute and, if the chamber containing the specimen is previously flushed out with nitrogen, this atmosphere can be maintained by a slow continuous stream of the gas introduced at a distance above the specimen. If this is sufficiently slow, it has no appreciable effect on the weighings but it may be turned off while they are performed. The additional advantage accrues that the $\kappa_2 v$ correction is now negligible (κ_2 for nitrogen $\simeq -0.0004 \times 10^{-6}$) and equation (3) reduces to

$$\chi = \left(\frac{2gl}{WH^2}\right)w \tag{4}$$

In addition to the calibrations of field and specimen length, already described, a temperature calibration is now needed. Probably the most satisfactory method, since it reproduces exactly the conditions of the measurements, is to use a substance for which the variation of susceptibility with temperature is known. $(NH_4)Fe(SO_4)_2 . 12H_2O$ obeys the Curie law and $CuSO_4 . 5H_2O$ obeys the Curie-Weiss law with $\theta = 0.7°$. The calibrant is packed into a normal specimen tube and w measured at intervals of between 5 and 10°. Putting $(2gl/WH^2) = C_1$ and combining equation (4) with the Curie-Weiss law:

$$\frac{C}{T + \theta} = C_1 w - d$$

where d is the diamagnetic correction per gram for the ligands of the calibrant.

$$\therefore \quad \frac{C/C_1}{(T + \theta)} = w - \frac{d}{C_1}$$

If just one value of T (along with the corresponding w) is accurately known, the constant C/C_1 is easily obtained and the temperatures calculated for all the other measured pulls. This eliminates packing errors which are introduced if reported values for C are used (indeed these values vary so widely for $CuSO_4 5H_2O$ as to be quite unacceptable for calibration purposes). In addition, errors in the field calibration are of minor importance since they affect only the relatively small constant, d/C_1.

If a thermocouple is being used to measure temperature, and an ice/water bath surrounds the "cold" junction, then the reference temperature may be taken as $0°$ C, since at this point the thermocouple e.m.f. is zero and the corresponding pull is readily obtained graphically. Alternatively, w can be measured at room temperature in the absence of any coolant. As the temperature is then completely uniform, this allows for the small temperature gradients occurring when coolant is present.

Temperatures above room temperature can be attained with a cryostat of the electrically heated type, if the refrigerant is dispensed with. Unfortunately, if the specimen is suspended below the balance, even a moderate rise in temperature creates convection currents which disturb the balance. Unless the whole system is modified and the specimen mounted above the balance, it must be accepted that measurements above room temperature are appreciably less accurate than those below.

Some typical results obtained with a Gouy balance at room temperature, and using an atmosphere of nitrogen, are given below.

A perspex tube was used. Its weight with the attached collar was 3.06328 g and, in a field of 6 210 oersteds, w_t, the apparent loss in weight was 0.00241 g. The tube was then filled to a height of 10 cm with powdered $CuSO_4 . 5H_2O$.

Weight of filled tube with field applied	5·19370
Weight of filled tube without field	5·17125
$\therefore\ w_{t+s}$	0·02245
w_t	0·00241
$\therefore\ w$	0·02486 g

$$W = 5{\cdot}17125 - 3{\cdot}06328 = 2{\cdot}10797\ \text{g}$$

$$\therefore\ \text{from equation (4)}\quad \chi = \frac{2 \times 981 \times 10 \times 0{\cdot}02486}{2{\cdot}108 \times 6\,210 \times 6\,210} \simeq 6{\cdot}00 \times 10^{-6}$$

Because of packing difficulties the absolute accuracy of this value may be no better than 2% (corresponding to 1% for μ_e). However, in comparing it with the other measurements at different temperatures, the relative errors depend not on packing but on the accuracy of the weighings and the temperature calibration and on the reproducibility of the field. In this particular case they should amount to less than $\frac{1}{2}$% in χ (or $\frac{1}{4}$% in μ_e), but obviously these figures are critically dependent on, above all, the magnitude of w.

FARADAY METHOD

It can be seen from equation (1) that, if the volume is sufficiently small for $H(\partial H/\partial l)$ to be constant over the whole sample, there is no need to integrate as was required in the Gouy method. This condition is the basis of the Faraday method in which samples of only a few milligrams are used.

With the usual substitutions equation (1) becomes:

$$\chi = \frac{\kappa_2 v + \left(\dfrac{g}{H \cdot \dfrac{\partial H}{\partial l}}\right) w}{W}$$

Because of the much smaller sample and the fact that only a portion of the field is utilized, the resulting pulls are very much lower than with the Gouy method. More sensitive weighing techniques are necessary and quartz fibre torsion balances, or Sucksmith ring balances, are most widely used. The sample can be placed in a fused quartz bucket, of about 1 mm internal diameter, suspended from the balance. With the fibre torsion balance an accuracy in weighing of better than 10^{-7} g can be attained and, being a null method, it involves no movement of the specimen with respect to the field. A Sucksmith balance consists of a phosphor bronze ring, fixed at the top and with the specimen suspension attached to the bottom. The force exerted on the specimen by the field deforms the ring and the deformation is magnified optically about 150 times (Fig. 49). Because of the large magnification the actual movement of the sample can be limited to a negligible amount. The main drawback of this otherwise simple system is that the range of a given ring is limited and interchanging is often necessary.

For measurements at room temperature only, a quite small pole gap and diameter will suffice so that a small permanent or electromagnet is adequate. In order to make $H(\partial H/\partial l)$ constant over as large a volume as possible, and hence make the positioning of the sample less critical, variously shaped pole pieces have been designed. However, the need for this is considerably reduced and the sensitivity of the method increased if the magnet can be moved vertically from below the sample to a position above it. Balance readings are

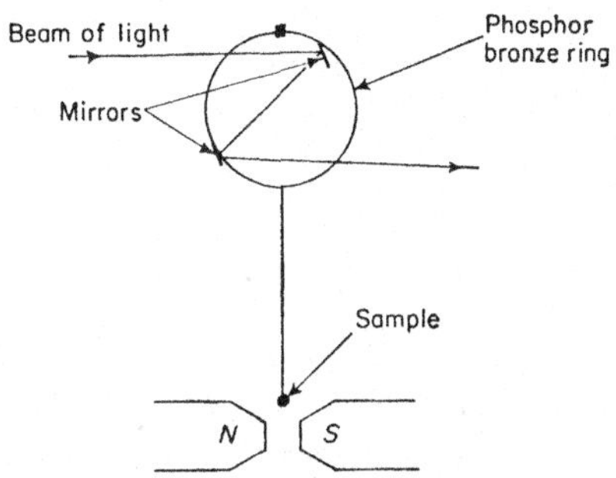

FIG. 49. Diagrammatic representation of a Sucksmith ring balance used in conjunction with the Faraday method.

then taken as the magnet is moved stepwise and range from a maximum apparent loss in weight to a maximum increase. The difference between these is then used to calculate χ.

The whole balance, suspension and sample system must be totally enclosed and may be flushed out with nitrogen to avoid the necessity of the $\kappa_2 v$ correction. The field is most conveniently calibrated by using a sample of known susceptibility.

If measurements over a range of temperature are required then, once again, a larger magnet is needed to accommodate the cryostat and to provide an adequate field. In general the dimensions are much the same as for the Gouy method.

The advantages of the Faraday method over the Gouy method are as follows.

1. A much smaller sample is needed. Apart from the question of availability of material this leads to the extra advantage that temperature gradients along the sample are quite negligible.

2. The accuracy is better. For moderate values of χ a sensitivity of 10^{-8} c.g.s. units, or better, is possible. This really follows from (1), since with such

D*

small samples the accuracy is no longer dependent on the density with which it is packed.

3. Ferromagnetic impurities may be corrected for, because the applied field does not vary significantly over the sample and is sufficiently high for the impurity to be approaching saturation.

4. Measurements above room temperature are convenient. The balances used are smaller than the conventional semi-micro balances usually employed in the Gouy method, and so it is possible to evacuate the whole balance, suspension and sample system. This removes the possibility of convection currents when the temperature is raised.

5. Measurements of magnetic anisotropy are feasible. By appropriately mounting a single crystal the susceptibilities along different axes may be determined. In this way detailed information can be derived about the presence and magnitudes of low symmetry components of the ligand field.

These advantages are however counterbalanced by a number of inherent disadvantages.

1. The equipment is generally less robust that the Gouy balance and requires more care in its use.

2. Samples must be very finely powdered. Although it is useful to be able to measure the magnetic anisotropy of crystals it is important, if the average susceptibility is to be measured, that no accidental alignment of crystals should occur. The fine powdering, necessary to avoid this possibility, presents the difficulty that co-ordination compounds frequently lose appreciable amounts of volatile ligands under these circumstances.

3. Measurements on solution are inconvenient. Though the method is, in principle, well suited to such measurements, difficulty arises due to losses by evaporation in the transference of very small amounts of liquid.

4. Impurities are dangerous. If the impurity is not dispersed homogeneously throughout the material, a small sample may contain a disproportionate amount. Careful grinding and mixing in bulk is a safeguard but scrupulous care must be taken to maintain the cleanliness of the sample, bucket and suspension.

QUINCKE METHOD

This is suitable for measurements on solutions and can also be used for gases. A column of liquid is subjected to a magnetic field which is homogeneous at the meniscus but falls to zero at the bottom (Fig. 50). It follows from equation (2) that the pressure, in dynes/cm^2, on the column is

$$P = \tfrac{1}{2}(\kappa_1 - \kappa_2)H^2$$

where 1 refers to the solution and 2 to the atmosphere above it. The level rises if the liquid is paramagnetic and falls if it is diamagnetic with respect to the gas above it. If the reservoir is sufficiently large to be essentially unaffected by changes (h) in the height of the liquid in the tube, then

$$h\rho g = \tfrac{1}{2}(\kappa_1 - \kappa_2)H^2$$

or

$$hg = \tfrac{1}{2}(\chi_1 - \chi_2)H^2$$

$$\therefore \quad \chi_1 = \frac{2hg}{H^2} - \chi_2$$

h may be measured to within 10^{-3} cm by a travelling microscope, but the sensitivity is improved by adopting a null method. This also makes the uniformity of the field near the meniscus less important. The meniscus is returned

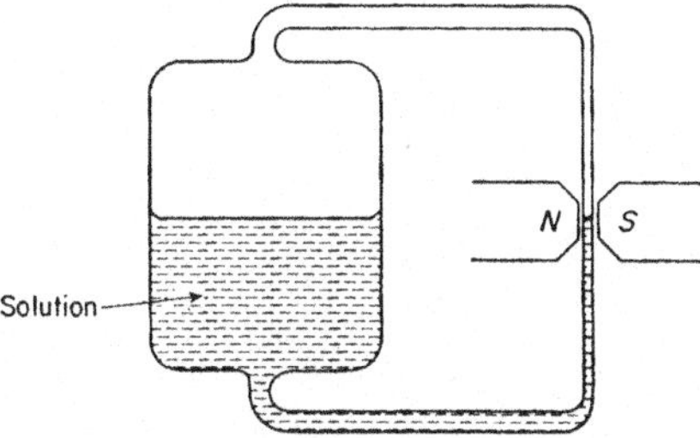

FIG. 50. Diagrammatic representation of the Quincke method.

to its original position, by changing the pressure of gas above it, its position being checked optically by a spot of light reflected from the surface.

A constant temperature jacket with water circulated from a thermostated bath will provide an effective temperature control.

Keeping the same atmosphere, the apparatus can be calibrated by using a liquid of known susceptibility. Alternatively, if the liquid is kept the same and the atmosphere changed, the susceptibilities of gases can be measured.

With dilute solutions this method is more accurate than the Gouy and also has the advantage of extreme simplicity, though with small diameter tubes care must be taken to keep the inside clean and so avoid surface tension effects.

NMR METHOD

Providing a nuclear magnetic resonance spectrometer is available, the procedure devised by Evans is very convenient.

The frequency at which proton resonance occurs depends on the magnetic environment of the proton and changes in this environment produce corresponding changes in the resonance frequency. If the environmental change is caused by the presence of a paramagnetic solute, it should be possible to relate the susceptibility of this to the change (Δf) in the resonance frequency of the proton. This is the basis of the Evans method in which the resonance frequency of a standard substance in a solution is compared to that of the same substance in an otherwise pure solvent.

The solution under investigation, along with 1–2% of the standard proton-containing substance, is placed in a capillary tube such that its length $\gg$ diameter. This is placed approximately coaxially inside a normal NMR tube containing the solvent with the same concentration of the standard substance

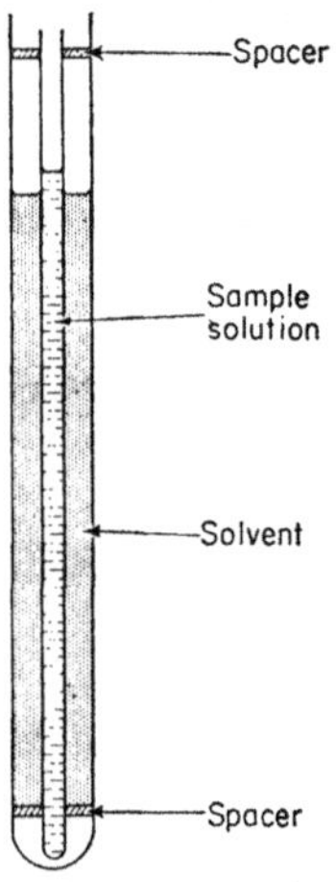

FIG. 51. The arrangement of the sample for the measurement of the susceptibility of a solution by NMR.

(Fig. 51). The tubes are spun as is usual during NMR measurements. Under these circumstances it can be shown that the susceptibility of the paramagnetic solute is

$$\chi = \frac{3}{2\pi m}\frac{\Delta f}{f} + \chi_0 + \frac{\chi_0\,(\rho_0 - \rho_s)}{m} \tag{5}$$

where f is the frequency of the proton resonance, n is the mass/ml of solute, and o refers to pure solvent and s to the solution.

t-Butyl alcohol is a convenient proton standard for aqueous solutions, while for non-aqueous solutions C_6H_6 or $(CH_3)_4\,Si$ are suitable.

This method is restricted to solutions but its sensitivity is good (differences in susceptibility of 10^{-9} c.g.s. units may be distinguished) and as little as 0·02 ml is required. With dilute solutions the last term in equation (5) is negligible. In fact it is best not to use concentrated solutions since, in these, the accuracy may be reduced by broadening of the resonance lines.

Magnetic Titrations

The experimental section would not be complete without some mention of the use of magnetic measurements in studying reactions as opposed to the examination of compounds which have been isolated.

If the addition of a reagent to a particular compound causes a change in the number of unpaired electrons, the course of the reaction can obviously be followed magnetically. If the reaction is performed in solution, samples may be extracted after each addition of fixed amounts of the reagent and the susceptibility measured. Corrections must be made at each stage to allow for the

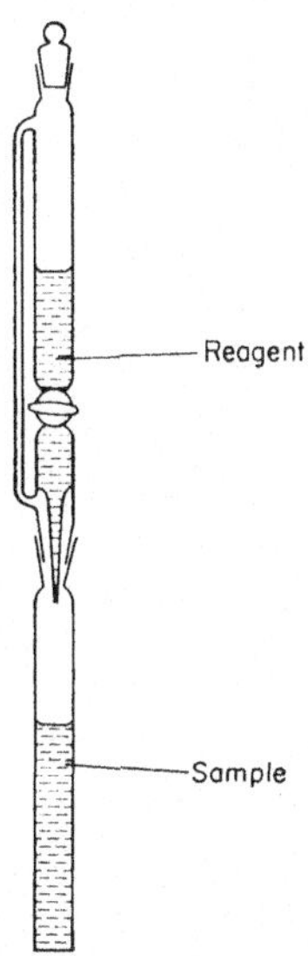

Fig. 52. A convenient assembly for a magnetic titration.

accompanying effect of dilution. This is a cumbersome and rather inconvenient procedure and is greatly simplified by the use of apparatus in which the burette, reagents, and reaction vessel are combined, the reaction vessel acting as the sample tube. Using the Gouy method, the whole assembly may be suspended from the balance (Fig. 52) so that the total weight does not vary and there is no need to repeatedly extract samples. Of course, the effect of adding the reagent is not only to dilute the reactants but also to alter the field gradient acting over them. The correction needed to allow for this can be obtained by a "blank" titration in which a second aliquot of solution is diluted in the same manner but with pure solvent.

The corrected values of w are then plotted against moles of reagent added, when a change in slope indicates the formation of a compound with different moment. From these data it is possible to calculate the formation constants of intermediate complexes, providing that they have moments differing (perhaps only in their orbital contributions) from the other species present.

This technique has been applied, for instance, to the reaction of Fe^{2+} with phenanthroline giving mono and tris complexes.

INTERPRETATION

When a result is obtained which merely confirms suspicions already held, there is obviously no difficulty in interpretation. However, one of the most fascinating, if at times infuriating, facets of chemistry is the frequency with which unexpected results turn up. When this happens it is necessary to examine carefully all the possible causes. For this reason it may be useful to reverse the procedure so far adopted and to look in turn at the different types of behaviour found experimentally, listing their possible causes, even though these may have no logical connexion.

In order to facilitate the understanding of experimental results it is useful to plot, as a matter of course, graphs of $1/\chi_A$ against T and also μ_e against T. Aspects of the behaviour which may be obvious on one are not always apparent on the other.

1. Plots of $1/\chi_A$ versus T

(a) *Linear*; $\theta = 0$

This is Curie law behaviour. The theoretical derivation of the Curie law is based on the simple assumption that the paramagnetic ions in a substance are associated with magnetic moments which are independent of temperature. This behaviour arises when the ion's ground level is a singlet or a non-magnetic doublet. It is the situation expected in magnetically dilute compounds of transition metal ions with S ground terms, or with doublet ground terms arising from D terms, or with singlet ground terms arising from F terms.

(b) *Linear*; *finite value of* θ

This is Curie-Weiss law behaviour. When exchange interaction occurs between neighbouring ions, the Curie-Weiss law holds at temperatures much higher than the Curie or Neel points. In magnetically concentrated systems a positive value of θ indicates antiferromagnetism and a negative value, ferromagnetism. In magnetically dilute systems intramolecular antiferromagnetism leads to positive values of θ, but the less common negative values are unlikely to arise from exchange interactions. Where it is known that antiferromagnetism is responsible for deviations from the Curie law it is permissible, providing that $T \gg \theta$, to use the formula

$$\mu_e = 2 \cdot 83 \sqrt{\chi_A (T + \theta)}$$

In this way it is possible to obtain a value of μ_e which excludes the effect of the interaction. At lower temperatures, though the Curie-Weiss law may appear to hold over a limited range, the behaviour becomes increasingly complicated and values of θ obtained by extrapolation cease to have significance.

There are vast numbers of compounds which give linear plots of $1/\chi_A$ against T and positive values of θ. It cannot, however, be assumed that this indicates antiferromagnetic exchange. The apparent applicability of equation (6) is frequently fortuitous and its use should always be viewed with caution.

The Curie law breaks down and may be replaced by the Curie-Weiss law for ions in which the ground term is degenerate or is associated with a higher level sufficiently close to be thermally populated. In such cases the resultant moment is a combination of the moments associated with the individual levels and is therefore temperature dependent. Interpretation resolves itself into deciding what influences are responsible for splitting an originally degenerate level, or alternatively for lowering an originally much higher level. Spin-orbit coupling is probably the most common cause but other effects, associated with the ligand field and leading to fields of lower symmetry or to spin paired–spin free equilibria, can also be responsible.

In ions which possess a triplet ground term (i.e. those for which an orbital contribution might be expected) spin-orbit coupling may split this term by an amount sufficient for the resulting levels to be populated unequally. This often leads to a positive value of θ which is without significance and in such a case attention is best restricted to the moment itself. Above about 200° K, $K_3[Mn(CN)_6]$ appears to follow the Curie-Weiss law with $\theta \simeq 100°$ and a moment of 3·5 B.M. at 300° K. This has been shown to be due to the d_ε^4 configuration under the influence of spin-orbit coupling with the possibility of a low symmetry field. The use of equation (6) here would be quite irrelevant.

(c) Non-linear

When this behaviour is associated with a moment below $\mu_{s.o.}$ it is usually due to antiferromagnetic interaction. The usual procedure is to determine, by whatever means are available, the extent of polymerization of the complex and then to try to fit the experimentally observed susceptibilities to the appropriate theoretical equation.

(d) Non-linear only at higher temperatures

For ions, particularly those with only one unpaired electron where the values of χ_A are relatively small, temperature independent contributions may be significant.

Where the contribution is t.i.p. the effect is to give values of χ_A higher than would otherwise be the case. This usually only manifests itself at higher temperatures as illustrated in Fig. 53.

Being negative, a diamagnetic contribution has the opposite effect causing the $1/\chi_A$ versus T plot to curve upwards at higher temperatures. Such contributions usually arise from the presence of diamagnetic impurity. Excess of uncombined ligand should be analytically detectable but, where a complex is prepared from a diamagnetic compound of the metal, appreciable quantities of this may be present without significantly affecting all the analytical figures.

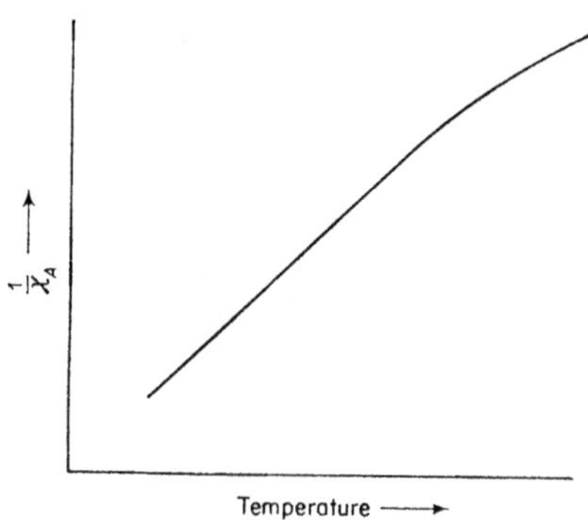

FIG. 53. A plot of $1/\chi_A$ against T showing the effect of t.i.p. at high temperatures.

(e) *Complicated non-linear*

If a particular material is in fact a mixture of compounds with quite different magnetic properties, a variety of behaviours may emerge. In the great majority of cases, of course, this would be recognized without the need for a magnetic measurement. That this is not always so, however, is well illustrated by the case of copper p-nitrobenzoate. Simple precipitation from aqueous solutions of cupric sulphate and sodium p-nitrobenzoate yields a product which analyses well for $Cu(NO_2.C_6H_4.COO)_2.H_2O$. Figure 54 shows the results obtained for a sample prepared in this way. This behaviour is in fact due to the presence of two different forms of the compound. One is almost certainly binuclear and shows appreciable antiferromagnetic interaction, while the other is probably polynuclear and shows only a small amount of interaction. The two forms are analytically identical but some separation may be effected since, on standing in the mother liquor, the former gives well-defined crystals while the latter is apparently amorphous. Figure 55 shows in a somewhat idealized manner how the initially determined values of χ_A may arise from the two quite distinct component sets of values.

2. PLOTS OF μ_e VERSUS T

(a) *μ_e independent of T*

This is case 1 (a) above and is Curie law behaviour.

(b) *μ_e varying with T*

As explained under 1 (b), this may be due to exchange interaction but apparent applicability of equation (6) must not be taken as confirmation of this. The action of spin-orbit coupling on a ground T term may lead to appreciable temperature dependence. In such cases, even if the theoretical values of

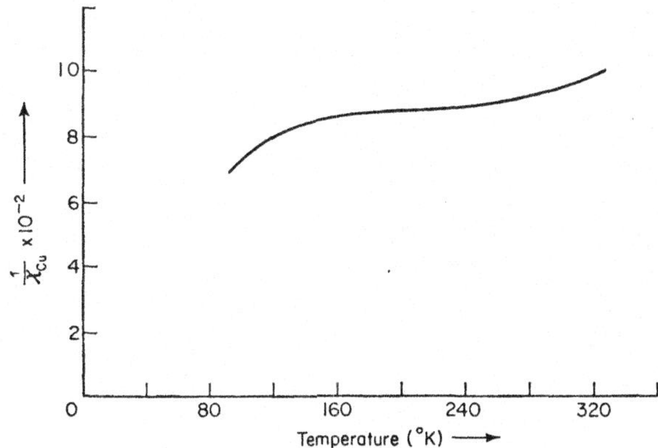

FIG. 54. A plot of $1/\chi_A$ against T for a sample of copper *p*-nitrobenzoate.

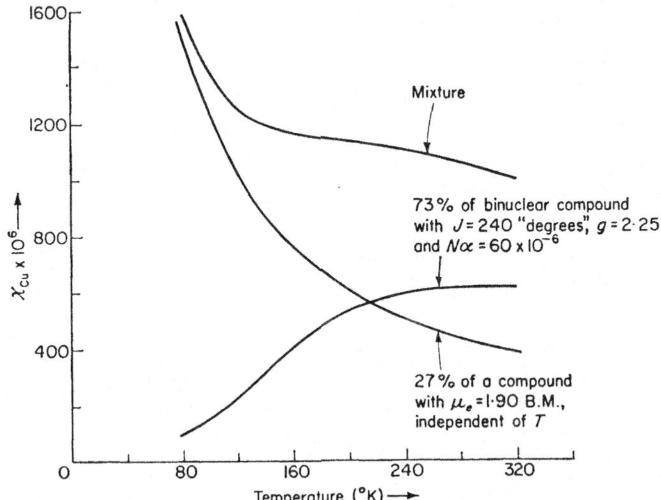

FIG. 55. Plots of χ_A against T for copper *p*-nitrobenzoate showing, in an idealized way, how the experimental results could arise from a mixture of a binuclear form (showing exchange interaction) and a form obeying the Curie-Weiss law.

μ_e agree only imperfectly with the experimental, it may be that asymmetric fields or electron delocalization are responsible for the difference. The behaviour of the previously mentioned compound, $K_3[Mn(CN)_6]$, may be explained, assuming perfect octahedral symmetry, only by using an improbably low value for the spin-orbit coupling constant. It seems more likely that a small, low symmetry, component of the ligand field is present.

(c) μ_e very low (<1 B.M.)

A room temperature moment of 1 B.M. corresponds to an atomic susceptibility of only about 400×10^{-6}. Such values are frequently t.i.p., particularly for second and third row ions. The smallness of these moments invariably means that the experimental errors involved in their measurement are unavoidably high, and slight changes should be viewed with suspicion. For instance, an error of say 20×10^{-6}, which is very easily introduced in the diamagnetic correction for large ligands, produces 2·5% and 10% errors in room temperature moments of 1 and 0·5 B.M. respectively, and can easily lead to quite erroneous conclusions about the temperature dependence of μ_e.

Perhaps even more important is the dominant effect of trace amounts of paramagnetic impurities. These may be analytically undetectable but their contribution to the measured χ_A or μ_e, particularly at the lower temperatures, may swamp that of the bulk of the material. Chromous acetate is a case in point. Its room temperature moment has been variously reported as being about 0·5 B.M. In view of the likelihood of the presence of traces of chromic ion, it has been suggested that the moment is entirely due to the impurity, the chromous acetate itself being diamagnetic. However, measurements at different temperatures reveal the behaviour shown in Fig. 56. This almost certainly arises from antiferromagnetic behaviour, due to strong interaction in the binuclear acetate, with the paramagnetic effect of the chromic impurity superimposed and becoming dominant at low temperatures. Figure 57 shows, again in an idealized manner, how the experimentally determined susceptibility may arise in this way.

(d) Abrupt changes in μ_e (and χ_A)

These are occasionally found but are not well documented and the origins of the phenomena are by no means clear. The compound formally represented as [Fe(phen)$_2$(SCN)$_2$] has a moment of about 5·2 B.M. and obeys the Curie-Weiss law down to about 180° K. At this point μ_e drops dramatically. Within experimental error the change is reproducible and appears to correspond to a transition from a spin-free to a spin-paired form. The residual paramagnetism at low temperatures is the only factor which varies appreciably from preparation to preparation and is presumably due to traces of impurity.

Another example is the nitrosyl of bis(salicylaldehyde)ethylenediimine iron (II). Above 190° K it follows the Curie-Weiss law with $\theta \simeq 100°$ and $\mu_{300} \simeq 3·6$ B.M. This is consistent with three unpaired electrons and appreciable antiferromagnetic exchange. The abrupt change then produces a moment of 2·1 B.M. below about 180° K which varies little as the temperature falls.

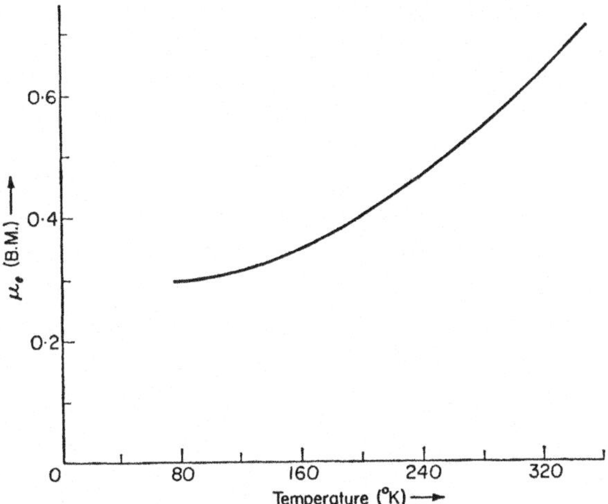

FIG. 56. A plot of μ_e against T for $Cr_2(CH_3.COO)_4\,2H_2O$.

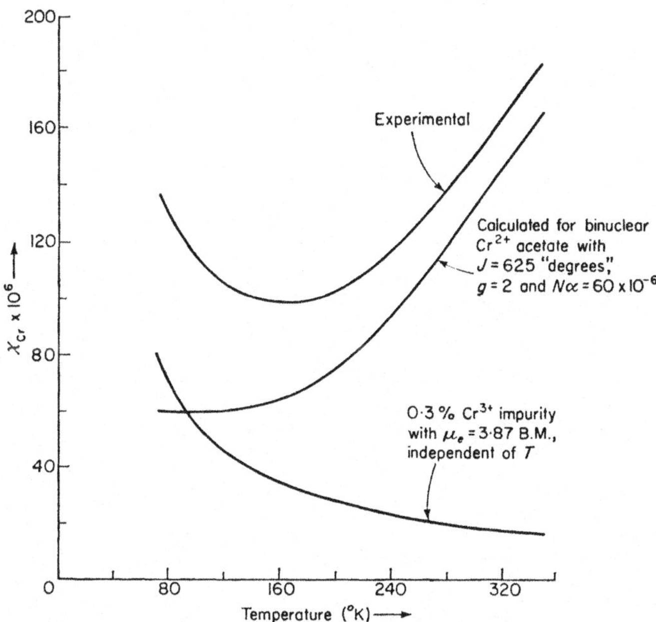

FIG. 57.   Plots of χ_A against T for chromous acetate showing, in an idealized way, how the experimental results could arise for a binuclear structure (with strong exchange interaction) and a trace of chromic impurity. For simplicity the chromic impurity is assumed to obey the Curie law, though in fact this is most unlikely.

This suggests one unpaired electron with little or no exchange. An interesting, and possibly significant, difference between the phenanthroline and nitrosyl complexes is that in the latter case the position of the sudden change in μ_e is a few degrees higher when the temperature is being increased than when it is being decreased.

It is obvious from the foregoing discussions that it is desirable to check magnetic determinations not only by repetition (particularly with increasing as well as decreasing temperature) but also by using samples from different preparations. In this way the effects of impurities, which are unlikely to be present in exactly reproducible quantities, may be detected.*

* For references for this chapter, see groups 8 and 11 on pp. 111 and 112.

APPENDIX

ENERGY

In view of the fundamental importance of energy in all processes of change, many different units are used, each appropriate to a particular field. It might therefore be useful to note some of the conversion factors of these units and to compare the magnitudes of some of the quantities regularly used by chemists.

WAVENUMBER

According to Planck, the energy of a single photon radiated at a frequency of v cycles/sec is given by

$$E = h v$$

where h is Planck's constant $= 6 \cdot 6256 \times 10^{-27}$ erg sec.

However, $v = c/\lambda$ where c is the velocity of electromagnetic radiation ($= 2 \cdot 9979 \times 10^{10}$ cm/sec in vacuum) and λ is the wavelength in cm. Rather than v it is often convenient to use $\bar{v}$, measured in cm^{-1} or wavenumbers

$$\bar{v} = \frac{1}{\lambda} = \frac{v}{c}$$

$$\therefore \quad E = hc\bar{v}$$

Thus the energy of a single photon radiated at 1 cm^{-1}

$$= 6 \cdot 6256 \times 10^{-27} \times 2 \cdot 9979 \times 10^{10} \times 1$$

$$= 1 \cdot 9863 \times 10^{-16} \text{ ergs}$$

This is an extremely small amount of energy but, if referred to a gram molecule of material in which each molecule radiates one quantum of energy, the total energy is then

$$hc\bar{v} N = 1 \cdot 9863 \times 10^{-16} \times 6 \cdot 0225 \times 10^{23}$$

$$= 1 \cdot 1963 \times 10^{8} \text{ ergs}$$

$$= 2 \cdot 8592 \text{ cal}$$

which, of course, is sufficient to raise the temperature of 1 g of water by approximately the same number of degrees.

Though the wavenumber is obviously not a unit of energy it is conventionally used as if it were, particularly in connexion with spectroscopic measurements.

Degree Absolute

kT is another frequently used energy term where k, Boltzmann's constant, $= 1{\cdot}3805 \times 10^{-16}$ ergs/degree. It is not uncommon to use the "degree absolute" or simply the "degree" as an energy equivalent

$$1 \text{ degree} = 1{\cdot}3805 \times 10^{-16} \text{ ergs.}$$

Conversion Factors

The more useful conversion factors may be obtained from the following list:

$$
\begin{aligned}
1 \text{ cal} &= 4{\cdot}1840 \times 10^{7} \text{ ergs} \\
&= 4{\cdot}1840 \text{ absolute joules} \\
&= 1{\cdot}1620 \times 10^{-6} \text{ kWh} \\
&= 2{\cdot}6117 \times 10^{19} \text{ eV} \\
&= 3{\cdot}0308 \times 10^{23} \text{ degrees absolute} \\
&\equiv 0{\cdot}34975 \text{ cm}^{-1} \text{ per mole of material}
\end{aligned}
$$

It follows that

$$
\begin{aligned}
1 \text{ kcal/mole} &= \frac{3{\cdot}0308 \times 10^{23}}{6{\cdot}0225 \times 10^{23}} \times 10^{3} \\
&= 503{\cdot}24 \text{ degrees absolute} \\
&\equiv 349{\cdot}75 \text{ cm}^{-1}
\end{aligned}
$$

Effect of a Magnetic Field

If an ion with a single unpaired electron is subjected to a magnetic field, its energy will depend on whether the unpaired electron is aligned parallel or antiparallel to the field. Energetically the difference between these two levels is $g\beta H$. If, for simplicity, the orbital contribution is assumed to be zero and the field is taken as 10 000 oersteds, this energy is

$$2 \times \frac{0{\cdot}92732 \times 10^{-20}}{1{\cdot}9863 \times 10^{-16}} \times 10000 = 0{\cdot}9337 \simeq 1 \text{ cm}^{-1}$$

Thus for a large number of such ions at room temperature, where $kT \simeq 200$ cm^{-1}, the ratio of the number of ions in the excited (antiparallel) level to the number in the lower (parallel) level is

$$e-\frac{gBH}{kT} \simeq e-\frac{1}{200}$$

i.e. the excess of ions in the lower level is only about $\frac{1}{4}\%$. Alternatively, to attain the separation of levels sufficient to force say 90% of the ions into the lower level would require, in the above case, a field of over $4\frac{1}{2}$ million oersteds. Since magnets generally used for susceptibility measurements have fields of the order of 5 000–10 000 oersteds, it is obvious that saturation effects may be neglected for all normal paramagnetic substances unless the temperature is close to absolute zero.

COMPARATIVE ENERGY VALUES

It is instructive to compare the orders of magnitude of different energy terms. The following is a list of quantities mentioned in the text with others commonly encountered. Where a range is indicated this is not intended to imply that values outside the range are not known but merely that the majority of cases lie within these approximate limits.

For transition metal ions:

Energy of spin coupling	$\sim$ 20 000–30 000 cm^{-1}
Energy of orbital coupling	10 000 cm^{-1}
Energy of spin-orbit coupling	100–1 000 cm^{-1} (first row)
Magnetic field splitting	1 cm^{-1}
Octahedral ligand field (Δ_0)	10 000 to 30 000 cm^{-1}
Lattice energies	200 000 cm^{-1} (600 kcal/mole)

Bond energies are usually of the order of 17 000–35 000 cm^{-1} (50–100 kcal/mole), and the hydrogen bond is around 1 700 cm^{-1} (5 kcal/mole).

REFERENCES

1. Bates, L. F. (1961). "Modern Magnetism", 4th ed. Cambridge University Press, London.

   Bleaney, B. I. and Bleaney, B. (1965). "Electricity and Magnetism". Clarendon Press, Oxford.

   Semat, H. (1962). "Introduction to Atomic and Nuclear Physics", 4th ed. Chapman & Hall, London.

These are a selection of the many available texts on the fundamental physical theory introduced in Chapter II.

2. Ballhausen, C. J. (1962). "Introduction to Ligand Field Theory". McGraw-Hill, New York.

An account of crystal and ligand field theories for the advanced student, requiring a basic understanding of quantum mechanics. It contains an extensive list of original references.

3. Cartmell, E. and Fowles, G. W. A. (1966). "Valency and Molecular Structure", 3rd ed. Butterworths, London.

Written for the undergraduate, this popular book provides a useful introduction to the study of molecular structure, of which magnetochemistry is an integral part.

4. Cotton, F. A. and Wilkinson, G. (1966). "Advanced Inorganic Chemistry", 2nd ed. Interscience, New York.

A comprehensive text on inorganic chemistry with an excellent section on transition elements in which crystal and ligand field theories are introduced.

5. Figgis, B. N. (1966). "Introduction to Ligand Fields". Interscience, New York.

A good account of crystal and ligand field theories containing a section in which the mathematical techniques of group theory, necessary for a full understanding of the book, are developed. References are confined to advanced textbooks and reviews.

6. Figgis, B. N. and Lewis, J. (1960). *In* "Modern Coordination Chemistry" (J. Lewis and R. G. Wilkins, eds.), pp. 400–487. Interscience, New York.

The section by Figgis and Lewis on magnetochemistry has been largely superseded by later reviews by these authors. Although now somewhat out of date, it has a useful summary of magnetic data on second and third row transition elements.

7. Figgis, B. N. and Lewis, J. (1964). *In* "Progress in Inorganic Chemistry" (F. A. Cotton, ed.), Vol. 6, pp. 37–240. Interscience, New York.

A review of the magnetic properties of transition metal compounds with an extensive list of original references. It contains a most useful critical survey of magnetic data invaluable to any research worker in the field.

8. Figgis, B. N. and Lewis, J. (1965). *In* "Techniques of Inorganic Chemistry" (H. B. Jonassen and A. Weissberger, eds.), Vol. 4, pp. 137–248. Interscience, New York.

A discussion of the magnetic properties of transition and inner transition metal compounds laying particular emphasis on experimental techniques. It contains a critical survey of experimental data on lanthanides and actinides.

9. Herzberg, G. (1944). "Atomic Spectra and Atomic Structure", 2nd ed. Dover Publications, New York.

An excellent book showing in a simple manner how the details of atomic structure are derived from atomic spectra. The first edition was published in 1937 but it is still justifiably regarded as one of the best on the subject.

10. Lewis, J. (1962). *Sci. Prog.* **50**, 419.

A good introductory account of the magnetic properties of first row transition elements, suitable for undergraduates.

11. Mulay, L. N. (1967). "Magnetic Susceptibilities". Interscience, New York.

This reprint of a previous article is a general account of the uses of magnetic measurements. A series of articles and reviews by the author and his wife have dealt extensively with the experimental applications of magnetic measurements usually with full details of the instrumentation involved.

12. Nyholm, R. S. (1956). "10e Conseil de l'Institut International de Chimie. Solvay", Stoops, Brussels.
Nyholm, R. S. (1958). *J. inorg. nuclear Chem.* **8**, 401.

Introductory reviews of magnetochemistry very useful at an undergraduate level.

13. Selwood, P. W. (1956). "Magnetochemistry", 2nd ed. Interscience, New York.

A survey of experimental techniques, applications and data of the whole field up to 1956. Interpretation is exclusively in terms of valence bond theory. Now out of date but still useful as a comprehensive starting point in the subject.

14. Van Vleck, J. H. (1932). "Electric and Magnetic Susceptibilities". Oxford University Press, London.

This is the book which laid the foundations of modern magnetochemistry. Much of it is too advanced for the average undergraduate chemist but the lucidity of its presentation makes it invaluable for anyone studying the development of the subject.

Two extensive but non-critical surveys of magnetic data have been compiled:

Foëx, G., Gorter, C. J. and Smits, L. J. (1957). "Constantes Selectionnées 7, Diamagnétisme et Paramagnétisme relaxation paramagnétique". Masson, Paris.

The literature of all compounds is surveyed up to 1955.

König, E. (1966). "Magnetic Properties of Coordinated and Organometallic Transition Metal Compounds", Vol. 2, Group II of Landolt-Börnstein (K.-H. Hellwege, ed.). Springer-Verlag, Berlin.

The literature is surveyed to the beginning of 1965.

Printed in Great Britain
by Amazon

53557439R00071